ARCHEOPSYCHOLOGY AND THE MODERN MIND

by Douglas Keith Candland

eBook version

ISBN: 978-1-105-19985-1

About the Book

You read an article about repressed memories of sexual abuse returning in middle-age; a television program features actors as villains who have a certain build and physiognomy; you chat with a friend about the damage done to their personalities by their parents, siblings, or circumstances; you explain to someone how you forgot a task assigned because of an unconscious motivation. We are all natural psychologists, explaining behavior by the beliefs of our time and culture.

We are captives, in a psychological sense, by theories and ideas that we accept tacitly, without knowledge or evaluation of their origins. We do not escape their influence, for they represent our idea of common sense. We can be, however, better evaluators of ourself and others by examining their origins and source of power.

To find and judge their source we are powerless if we use only our own sense of reason, for reason is contaminated by ideas of the past. Our best hope is to search for their origins as one would search to examine earlier civilizations by examining their fossils and remnants, by attempting an *archeopsychology*.

This book explores the idea of an *archeopsychology*; a term borrowed from historians of the human intellect, an idea that expresses an attempt to uncover and evaluate remnants of cultural ideas to show that they re-appear in modern thought.

The core ideas selected are (1) the mind can escape itself (as shown by hypnotism, animal magnetism, some modern metaphysics); (2) phrenology, that areas of the body, especially the brain, correspond in size or anatomy to behavioral traits (phrenology, some neuroscience); (3) evolution or a supreme power favors human progress; (4) body-type (sanguine, florid, thin, fat) corresponds to temperament; (5) body-type predicts the probability of criminal behavior (constitutional psychology); and, containing elements of all of the above, (6) selection from the gene pool of favored characteristics (eugenics, some extensions of sociobiology) to promote human progress and well-being.

The tone does not demean ideas of other times and places, but attempts to place these ideas as serious attempts to grasp the workings of the mind. The work is an intellectual history of these six ideas, each selected because they represent rich examples of how psychology reinvents longstanding core ideas.

What the book is not. It is not a text of names and dates in the history of psychology. It is an exploration of the intellectual history of six chosen topics. It is not pejorative or a book that ridicules previously-held ideas. Rather, it sees these as data for an understanding of how the human mind clings to various ideas by re-inventing them. It is not a work that tells the reader *what* to think. Rather, it offers the reader a partnership with the author in understanding the longstanding metaphors that guide our understandings of behavior and the mind.

About the Author

Born in Long Beach, California, USA, in 1934, the author received the AB degree from Pomona College and PhD from Princeton University. He has held postdoctoral and visiting professorships at the University of Virginia, the Pennsylvania State University, the Delta Primate Center of Tulane University, the University of Stirling (Scotland), the Medical Research Council, Cambridge (UK), the University of Mysore (India), and the University of California at Berkeley.

He is the author of *Psychology: the experimental approach* (McGrawHill, 1968, 1972), *Emotion* [with others] (Brooks/Cole, now Thompson, 1973, 2003), and *Feral Children and Clever Animals* (Oxford, 1993,1995). The author has appeared as a commentator on matters of psychological interest on BBC, CBS, National Geographic, Russian State television, and French television. He is the recipient of the American Psychological Foundation award for Distinguished Teaching and a like award from the Animal Behavior Society of America.

Now Professor of Psychology and Animal Behavior (Homer P. Rainey Professor, emeritus) at Bucknell University, where he taught from 1960 to 2002, he is the current editor of the *Review of General Psychology*.

TABLE OF CONTENTS

Chapter 3: The Congress of Ideas

Chapter 4: Unseen Powers: Spiritism and Other Worlds

Chapter 5: The Mental Fossil of Physiognomy; Character and Talent

Chapter 6: The Mental Fossil of Physique and Madness

Chapter 7: The Mental Fossil of Predicting and Treating Criminal

Chapter 8: The Mental Fossil: 'Human Progress' through Eugenics

Chapter 9: Revisiting Archeopsychology

Appendices

Chapter 1

The Archeopsychology of Mental Fossils

Cognitive science often carries on as though humans had no culture, no significant variability, and no history. - Merlin Donald, Origins of the Modern Mind

[It is] *highly probable that with mankind the intellectual faculties have been mainly and gradually perfected through 'natural selection' and this conclusion is sufficient for our purposes.* - Charles Darwin, Origin of Species

These words are an exercise in archeopsychology as it attempts to elucidate past minds from an archeological perspective. We start with the notion, if only as a hypothesis, that similar ideas appear and re-appear over time. To be sure, they are modified by culture, and therein lies our challenge, for the ideas may appear to be new and inventive to our contemporary culture when, upon inspection, we can find that they have appeared and re-appeared in cultures past. The social sciences in general and the study of psychology, in particular, are especially encumbered by ideas surfacing repeatedly, each time enhanced by our thinking them to be new. It is difficult to make progress when we are only arguing about the same ideas disguised by time and culture---and a lack of appreciation if what has gone before.

Douglas Keith Candland

Archeopsychology

That we may 'dig' for ideas, using as a tool a plan that has been suggested by intellectual historians who have proposed the term 'archeopsychology' for the venture. For example, the historians Le Goff and Nora in 1985 wrote:

"There is much talk of the history of mentalities, but convincing examples of such history are rare. It represents a new area of research, a trail to be blazed, and yet at the same time doubts are raised as to its scientific, conceptual, and epistemological validity. Fashion has seized upon it, and yet it seems already to have gone out of fashion. Should we revive or bury the history of Mentalities?" [1]

LeGoff presents the case that history should and must study mentality because he felt that, "the mentality of any one historical individual, however important [i.e., however unimportant], is precisely what that individual shares with other men of his time." Further, "The history of mentalities, then, represents a link with other disciplines within the human sciences, and the emergence of an area which traditional historiography refused to consider . . . The history of mentalities is to the history of ideas as the history of material culture is to economic theory" [2] High prospects indeed! Le Goff offers advise for the practice of an archeopsychology:

"What men say, whatever the tone in which they say it, conviction, emotion, bombast is more often than not simply an assemblage of ready-made ideas, commonplace and intellectual bric-a-brac, the remnants of cultures and mentalities belonging to different times and different places. This determines

2

the methods which the historian of mentalities must use.

"Two stages may be noted: first, the identification of different strata and fragments ["*striates et Bordeaux d'archeopsychologie*" in the original], this the maiden usage of the term of what, following Andre Varagnac's term 'archeo-civilisation', we may call 'archeopsychology' and, secondly, since these remnants are nevertheless ordered according to certain criteria of mental, if not logical, coherence, the historian must determine these psychic systems of organization . . ."[3]

Le Goff cites the 'mentalities' of Christopher Columbus as an example. However much Columbus's own mind supposedly differed from those of other people of his time, differed because of his willingness to argue for a western route to the treasures of Asia and the Orient, whatever else was in his mind was firmly set in his time and place. Imagine his notions about medicine, religion, law, or ethics. His ideas of these, his understanding of himself, were presumably much like those of his sailors. To understand Columbus's mind, we must come to understand the beliefs of the time. It is no news, of course, that all of us are trapped in the mentality of our times, but to know the nature of those mentalities is difficult, because our ways of thinking are locked into the orbit dictated by current mentalities. To perform archeopsychology, we must escape our own times: an impossible task. But, perhaps, was can extend our knowledge far enough to imagine how minds of former times understood their world. To do so requires thinking not so much about the history of psychology as commonly

3

practiced by scholars, but our inserting ourselves into the phenomenology of another time and place.

The possibility of an archeopsychology is a fundamental notion of Foucault's as well as of historians. [4] Although through him, the idea has achieved some attention from historians and sociologists, it has been resisted, if not wholly ignored, by psychologists---those who study the human mind. Both Freud and Jung showed interest in an archeology of the mind, although their followers have chosen not to highlight this interest, evidently thinking to be aberrant such interest. Freud's [5] works on the origins of humankind's symbolic thinking and his interest in Egyptology are evidence of his interest in the evolution of mentations while Jung's [6] emphasis on the mind's carrying archetypes, universal symbols of the mind akin to Platonic 'forms', surely offers a powerful statement regarding the possibility for an archeology of mentation.

What Columbus, Freud, and Jung share is a system if beliefs characteristic of folk if their time and place. Each reached beyond those constraints to think in different ways, thereby altering how folk of succeeding generations understood themselves. Nonetheless, their ideas---of geography, exploration, and the mind---can lead us to find the antecedents----the fossil remnants of earlier folk---that provide the core idea. Let us consider an example of how psychological thinking---that is, thinking about how our minds work---can be elucidated by archeopsychology. The example is the change in meaning of the term 'folk-psychology' and the faddish use of the term 'meme'.

Folk-psychology

The richest source of ideas that tempt the work of the archeopsychologist always lies in 'folk-psychology'. Some contemporary examples include the belief that body-type is related to temperament, that traits such as honesty and integrity can be foretold by facial expressions, that there are two worlds, the one real in which we dwell and the one unseen and spiritual, this a basis of all religions, that thee is a genetic component to criminals and the dishonest, and that different ethic groups display different, probably inborn, traits. Thinking about the folk-psychologies that we carry with us is the first step toward imaging how ideas came to be, this he first step for the archeopsychologist.

Wilhelm Wundt [1832-1920], the pioneer of experimental psychology and trainer of the first generation of German and American academic psychologists, uses the term "Völkerpsychologie" to refer to the fact that experimental psychology, while it may do well at measuring the conscious state of individuals, is not well suited for studying the consciousness of groups of individuals. His ten volume "Völkerpsychologie" [1900, 1906, 1914, 1916, 1922] and his *Elements of Folk Psychology* [1916] presented a clear use of the idea of group-consciousness, but the fact that all these volumes have not been translated into English has provided an opportunity for English-speaking academics to re-invent the term 'folk-psychology' and provide it with many meanings. It is this sort of confusion that archeopsychology can attempt to clarify.

Contemporary users of the term 'folk-psychology' would do well to examine Wundt's use of the term, and why it required ten volumes to make his point. He writes:

"The word 'Völkerpsychologie' [folk psychology] is a new compound in our [the German] language. It dates back scarcely farther than to about the middle of the nineteenth century. In the literature of this period, however, it appeared with two essentially different meanings. On the one hand, the term 'folk psychology' was applied to investigations concerning the relations which the intellectual, moral and other mental characteristics of people sustain to one another, as well as to studies concerning the influences of the characteristics upon the spirit of politics, art, and literature. The aim of this work was a characterization of peoples, and its greatest emphasis was placed on those cultural peoples whose civilization is of particular importance to us — the French, English, Germans, Americans, etc." [7]

Wundt then tells us that by folk psychology he does not mean 'community psychology' or 'social psychology', nor 'sociology', but, using my translation, a 'cultural psychology', a psychological-based history of humankind. He suggests four stages in the development of folk psychology: that of the primitive 'man', a totemic age in which objects were worshiped for their symbolic meaning, an age of heroes and gods, and, in his time [1910], the development of the national state and the national religion. The case for these four stages occupies the remaining 498 pages of his translated book.

The translator, Edward Schaub, a professor of philosophy at Northwestern at the time of the First World War, advises us that:

"It should not be overlooked, however, that the examination of the mental processes that underlie the various forms in which social experience comes to expression involves a procedure which supplements, in an important way, the traditional psychological methods. More than this, Wundt's Völkerpsychologie is the result of a conviction that there are certain mental phenomena which may not be interpreted satisfactorily by any psychology which restricts itself to the standpoint of individual consciousness. Fundamental to the conclusions of the present volume, therefore, is the assumption of the reality of collective minds." [8]

But Wundt's definition and understanding of 'folk-psychology' has become lost. Writing in a 1960 introductory psychology text, Calvin Hall [1909-1985] begins a discussion of 'folk-psychology' by saying that:

"This approach may also be called popular psychology because it consists of notions of the average person as to why men and women behave as they do. It is self-evident to many that 'sparing the rod spoils the child,' that 'slums breed crime'." [9]

In recent years, the term 'folk psychology' has been pushed further in that direction. Stitch writes:

"In our everyday feelings with one another we invoke a variety of commonsense psychological terms including 'believe,' 'remember,' 'feel,' 'think,' 'desire,' 'prefer,' 'imagine,' 'fear,' and others."

The use of such terms is governed by a loosely knit network of largely tacit principles, platitudes, and paradigms that constitute a sort of folk theory. Following

recent practice, I will call this network 'folk psychology'.
[10] I agree that the categorization, analysis, and
formalization of the concepts used by folk to explain their
and others psychology is a reasonable datum of
anthropology and psychology. An example from
contemporary psychology of a 'folk' psychology is the
concept of the 'meme' which, ironically, became a meme
of itself.

The Meme of the Meme

A recent innovation in thinking about the evolution of
ideas must be Richard Dawkins' invention of the term
"meme". In the simplest terms, if human ideas are to be
viewed as evolving, in the most extreme sense of the
metaphor, then the meme is the unit of selection,
equivalent to the gene in biological evolution. (Of course,
there is much debate over the role of genes in biology,
and especially with regards to evolution and natural
selection, but the assertion stands for what it is.) If we
can properly define a meme, presumably we can learn
how to investigate the course of its transmission or
extinction. Dawkins' notion of the meme leaves ideas as
simple, tiny units that exhibit the grace and agility of a
gazelle. Examples of memes are tunes, ideas, catch-
phrases, clothes, fashions, ways of making pots or of
building arches. Just as genes propagate themselves in
the gene pool by leaping from body to body via sperm and
eggs, so memes propagate themselves in the meme pool
by leaping from brain to brain [11] The idea of the meme,
the unit of 'idea' that culture may spread, transmit, or
terminate, which may thereby appear and disappear in
different generations is engaging. In fact, the term 'meme'
was introduced in a footnote, perhaps facetiously. The
idea of an idea being a simile to a virus that spreads

rapidly did just that, becoming a meme whose importance was attested to by books on the topic. Dawkins, in conversation, used the idea of the 'rally cap' as an example of a meme. Because a baseball team rallied when some players wore their hats backwards, soon all wore them in like fashion when a rally was needed. Fans saw this and did so. In time, so wearing the cap became a collegiate fashion style. In more time, young people worldwide, many who had never seen a baseball game, adopted the idea. And so a meme spreads, eventually bearing no relationship to its origins. Such is the origin and fate of many ideas we hold about the workings of our minds.

If we examine enough physical fossils, we may find that they have in common a backbone, an ear, or an optic tectum. Such structures are constantly altering, yet their similarity evidences (usually) common origin and the workings of similar developmental principles. So too, a budding study of archeopsychology or memetics might concern itself with finding the common origin of, and principles which lie hidden behind the surface-appearance of fleeting trends.

What archeopsychology promises is a demonstration of the evolution of ideas, especially of those that intend to describe the workings of the human mind. It does so by treating ideas as fossils, as evidences of ideas who become transformed over time and culture but which retain a central idea which, once uncovered, can be seen to be archetypal. Archeopsychology is a way of thinking that uses the historical development of ideas as the most valuable tool in locating the cores of our thinking about our own minds and those of others.

Douglas Keith Candland

The Mental Fossils Metaphor

Tracking Changes. I use the word 'fossil' not in its strict chemical and physical sense, but in a metaphorical sense. The physical fossil record may indicate relationships developed in the past, or it may, at any one time, reveal oddities whose significance are as yet unclear. We search for mental fossils whose structure allows us to follow changing ideas as they re-group throughout intellectual history.

The meaning of these mental fossils, as is true of the meaning of many physical fossils, will be uncertain and we can only speculate as to their significance. Discovery and re-examination of physical fossils is so frequent and quarrelsome as to suggest to some that this is no science at all. The repeated re-assessments of human evolution come to mind, for herein the discovery of even a tiny fossil remnant can and does lead to a thorough redefinition of human origins. Rarely is such so decided without disagreement. Such disagreement is both inevitable and encouraging, for it establishes the hypotheses that drive future research. So long as we remain in the discovery stage of the fossil record, mental or fossil, observers will differ as to what they expect a find to unveil.

Digging. The plan to uncover mental fossils means that we must dig. We must 'unearth' the systems of belief that were prevalent at identifiable times and places. As is true of physical digs, mental digs are layered. The archeologist and paleontologist have tools and theories. Our archeopsychology tools are books, journals, film, and any other record we can find of what people thought, believed, and felt emotionally. Like fossils, these data are so scarce that we have no idea how representative any may be. A set of bones may represent different species, or differently sized members of a species, or differences in sex. Only context allows the paleontologist to establish the identity. Mental fossils can be expected to be more difficult to interpret we have almost no established framework on which to build.

In our preliminary unearthing and comparing of mental fossils, we must expect the same turns of fate and disappointments as does the archeologist. We must remember that digs collapse. Sometimes great efforts — whole lifetimes and fortunes — are spent on excavations that turn up little of value after an initially prized discovery. Sometimes there is simply nothing worth the effort. Sometimes one gets head deep (in the ground or in the book stacks) only to find that environmental circumstances have wiped out all traces of the past. Both physical and mental evidence alike may be lost due to flood, poor preservation, or the high acid content of given age's paper.

Dressing and Redressing. A third aspect of the archeopsychology metaphor is that of having to discover similarities between things that appear different. Just as a variety of skeletal variations may belong to the same species, so a variety of seemingly disparate ideas may have an underlying commonality. Many of our fossil finds may be different dressings and re-dressings of the same of idea, in the same sense that a fossilized young calf and a fossilized full grow bull may be the same 'type' of animal. As when examining physical fossils, we should expect that, though the structures found at different times and places may appear different, when compared properly, we can see an orderly shifting in design. Identification of the underlying contents of recurring beliefs reveals that one aspect, then another, is altered in ways seemingly slight, yet, considered together, is of rich significance. At first glance, we see totally different structures, but on tidier and longer investigation we see the small changes that underlie and expose the mental equivalent of speciation.

Story-Telling. The final aspect of our metaphor is that the results of digs need significant arrangement before they become sensical. Fossils are uncovered haphazardly even within a given site, often the finds of several sites must be combined to approximate a full skeleton, and rarely are full skeletons are never found. Individual fossils must be ordered and cataloged, placed in many different arrangements, and hypotheses are generated and discarded to determine if a given arrangement has coherence. Reports on most digs, physical and mental, begin with a statement as to what the searchers hope to find, The report then tries to make coherent the results of its excavation and the implications of its finds. Such is our task in searching for mental fossils.

Four Major Themes

I hope, in this series of digs, to follow the path or opportunities of four major ideas. These exemplars all had an especially dazzling brilliance at the Chicago World's Fair of 1893, but they can be found if in a simpler structure long before. Following the fair they continue to receive many re-dressings, and though they often are not always found together at later sites, their tenacity and tendency to commingle is impressive. The ideas continue to intertwine into the present. What is today's news is often a merely redressing of a longtime theme. Modernization does not guarantee truth and often merely disguises the origins of the idea. Here are the four themes that I consider to be both longstanding and often redressed. Think of them as basic fossil structures that have undergone many redressings.

Variation. The first of these examples is the idea of 'variation'. I propose that only recently has 'variation' become a staple of 'folk psychology'. From nineteenth century Britain came intellectual contributions from Darwin, Wallace, and Galton, that would invert our way of thinking: the focus shifted, like that of a figure-ground relationship. Charles Darwin [1822-1911] and Alfred Russel Wallace [1813-1923], co-discoverers of the importance of natural-selection, understood variation to be the key to understanding how the living world came to be as it is. In particular, they showed that species were not to be thought of as fixed entities, but instead as a concept produced by and best described by contrasting variation within and between different groups of organisms. There was no perfect archetypal form of elk, except in the mind of the man; there is but variation on the central core of what we call elk.. Given how long the opposite view had been held, that species where characterized as approximations to a perfect form, it is not surprising that altering this view created a rippling effect that brought other variation into stern relief.

Everywhere focus shifted from the study of "essences" to the study of variation. Into this changing world waltzed Francis Galton [1822-1911], Darwin's half-cousin (they shared the same grandfather, Erasmus Darwin [1731-1802]), who understood how to measure and describe variation. While Darwin wrote of variation in a general and abstract way, often providing select (but thorough) examples of its scope, Galton measured it directly, even obsessively, and invented statistical methods for quantifying variation. Galton's measurement of human variation gave flower to the use of tests to establish differences in human abilities and the promotion of eugenics. (For this he is today widely derided but, as we shall see in Chapter 8, we have reason to be concerned about our own uses of human variation.)

Measurement of human differences in intelligence and ability were featured at the Chicago Fair of 1893 to be described in Chapters 2 and 3.. The measurements formed the first exhibits of the nascent science of psychology in America. There development would lead to nationalized tests, today used for admission to universities and professional schools, but once used to encourage 'appropriate' matings and reproductions, and, at times, by the sterilization, castration, and vasectomization of young persons regarded as genetically inferior. (Chapters 5 and 6)

Human Progress. The second example offered is that of 'human progress'. Though the idea has been ubiquitous in history, at least in Western society, at the close of the 19th century, it was advanced in ways unmatched by previous times in unanimity and intensity. Wallace, who in addition to being a scientist was an explorer and thoughtful commentator on intellectual and social issues, choose to write, in 1899, *The Wonderful Century*. In this book, he tells that, "A comparative estimate of the number and importance of [this century's] achievements leads to the conclusion that not only is our century superior to any that have gone before: it must therefore be held to constitute the beginning of a new era of human progress." Pointing out that speed of travel had not changed from Roman times to the 1830s, he marveled at the steamship and train. Photography, telegraphy, x-rays, and the principles of natural selection, all received praise for their revolutionary promise, [12] just as they did at the 1893 Fair. These are inventions that have in common the transmission of the unseen and unseeable. They represent, metaphorically, two worlds: one unobservable and one 'real' and useful. They are metaphorically like our human sense of different worlds, the one seeable, material, and knowable and the other spiritual, unseen, transcending the material. This I call 'spiritism' to distinguish it from the practice of 'spiritualism' which, by the way, is set next to 'psychology' in the classification of knowledge made by the Library of Congress. (Chapters 2 and 3)

Spiritism. The third example offered is the human belief in dual worlds, the one a world of 'real' sense perceptions, things that can be seen and touched, and the other a world unseen but powerful in directing our behavior. Even natural selection leading to variability (now seemingly the archnemesis of any other-worldliness) could be taken as evidence of an unseen power. This view was largely prompted by essayist Herbert Spencer, Darwin's neighbor and admirer, inventor and promoter of the phrase "survival of the fittest". Spencer is generally credited with promoting, if not developing the idea of 'social Darwinism' (be wary of friends who attach your name to their ideas). He thought that natural selection would lead to a better world and better species, and that such betterment was inherent in God's plan, and therefore, Good. Darwin, troubled by the word 'better' seems not to have been convinced. I suggest that the putative origin of experimental psychology, the invention of psychophysics, is an attempt to associate the worlds of spiritism and the material by experimentation. (Chapter 4)

Physiognomy. The fourth example is that differences between physical bodies represent differences between minds or between temperaments and abilities. Ideas about the nature of arise from the longstanding western tradition of separating mind from body, as exemplified by Descartes' notions. Attempts to measure such differences have taken several forms: measuring the skull, measuring "body-type", measuring nose and eye shape, and attempting to correlate such measures with mental illness, criminality, personality, feeblemindedness, and a myriad of other combinations. Although on the one hand, our intellectual culture dismisses these ideas as 19th century pre-science, the notion guides the lay interpretation (and sometimes the scientists' interpretation, or modern neuropsychology. (Chapters 5, 6, and 7)

Our Itinerary

To paleontology and physical anthropology, no physical fossil is irrelevant, for each will find its place. To archeopsychology, no action or idea is wrong or silly. Each is a remnant that when tied properly to its source and evolution is an explanation of who we are and how we came to be mentally what we are. In like fashion, no fossil is useless, although we may not understand its history or function. We store physical fossils carefully, knowing that in time its role will be postulated. So it should be with mental fossils. Above all, we must practice at suspending our judgments until the fossils are laid out for inspection.

This book take us to unexpected and unusual places: to the Midway featured during the Chicago Fair; the Phrenological Cabinet, a museum and retail shop in lower

Manhattan in New York City; various prisons, mental institutions, bath-houses, autopsy tables, and medical 'inns' in which castrations were performed to improve behavior; to a hypnotized tree in a village in France; a resort-hotel designed for the production of Aryan children; octagon houses in New England; to files on the genetic history of genes of Americans; to Indiana and Virginia, USA; to the US Supreme Court, where state laws were approved allowing vasectomies of delinquents and the castration of the feebleminded and insane; and beyond.

Armed with our simple tools, our search for ideas and their adventure through time begins. The tools — these of our ability to read and to understand our intellectual history — seem too course and simple for the task ahead, but we press ahead.

Chapter 2

Initial Excavations

We need to examine prospective places to dig. Some will prove promising enough that we shall return to them later. Others will help us decide the location of promising future digs. Still others will provide amusement and perhaps suggest good dig sights to those interested in species other than our chosen ones. Each of the brief, teasing, descriptions of sites that follow will benefit from the detailed analysis that follow in individual chapters.

Vienna, 1779: Young Mozart, Dr. Mesmer, and Fraulein Franzl

While visiting Vienna in 1779, staying with longtime family friends, the young musician Wolfgang Amadeus Mozart wrote to his father, who was in Salzburg at the time, the following:

"Dear Father, Can you guess from where I am writing this? In the garden of Mesmer's house in the Landstrasse. Frau Mesmer is not at home, but Fraulein Franzl, now Frau von Bosch, is here. On my honour I would not have recognized her, she is so fat and healthy, and she has three children!"[1]

Mozart was awed by the change in the person he had known during previous visits as Fraulein Franzl, for before she had been disturbed by many illnesses, including convulsions, vomiting, an inability to urinate, toothaches, ear aches, depression, trances, temporary blindness, paralysis, and hallucinations. The ministrations of physicians did little to heal, so she turned to Frau Mesmer, the husband of a close friend, Dr Mesmer.

Dr. Franz Anton Mesmer was a physician, practitioner, and researcher on the notion of a magnetic 'fluidum', an invisible spirit that held the universe and the individuals in it in place. When the natural flow of the fluidum was disrupted, he postulated, illness resulted. The task of the healer, he taught, was to reestablish the natural spirit by altering the magnetic field both inside and outside the sick person. Dr. Mesmer's treatment of Fraulein Franzl consisted of passing three magnets over her body. Two of them were shaped as splints and the third was shaped as a heart. He tells us:

"I finally decided to produce an artificial ebb and flow in the patient's body with the help of some magnets. . . I tied two magnets to her head and hung another, a heart-shaped one around her neck so that it touched her breast. Suddenly a hot piercing pain rose along her legs from her feet, and ended, with an intense spasm, on the upper rim of the iliac bone. Here this pain was united with an equally agonizing one which flowed from both sides of the breast . . . in turn, pains shot up to the head and united in the parting of the hair. When those pains ceased, the patient felt a burning sensation, like glowing coal, in all of her joints . . . This lasted throughout the night. The entire side of her body, which had been paralyzed during the last attack, perspired freely, and in this part of the body the pain gradually ceased." [2]

Repeated treatments were applied during the next five years, and when Mozart met her again, as he described in the letter to his father, he found her well, happy, fat, married to Mesmer's stepson, and three times a mother. What Dr. Mesmer did for Fraulein Franzl, he did for many others seeking his cure for a variety of ailments. Indeed, Dr. Mesmer's healings became so well-known, the demands on his time so great, and his attention so splintered, that he had little time to observe some of the 'side-effects' of his healings. One of these side effects, however, was noted with fascination by a follower of Mesmer, the Marquis de Puységur [1751-1825], and has become better known than Mesmer's intended effects.

France, 1784: Puységur Cures Victor Race

In 1784, five years after Mozart had written about Fraulein Franz, Victor Race, a 23 years old peasant, took himself to the estate-owner, the Marquis de Puységur, in the hope of being relieved of the burning within his lungs. The Marquis, eldest of three brothers from a family known for their military commands and land holdings, was fascinated by Dr. Mesmer's achievements. He thought the medical establishment had been much mistaken in its dismissal of Dr. Mesmer's ideas and practices. The Marquis was himself an important figure in the 'Society of Harmony', a more or less secret society that supported and continued Dr. Mesmer' s works after professional medical groups had attacked them and punished Mesmer. Puységur undertook for Victor the treatment called 'animal magnetism', a treatment he had learned from Dr. Mesmer.

An important characteristic of the treatment was that the patient sometimes underwent a 'crises', as it was called by the French, a period when the patient seemed unaware of her or his own consciousness or, at times, adopted another identity. Often, those who experienced the *crises* were then cured of their complaints, so the state of the crises was considered a positive step toward cure. However, instead of passing into the expected crises, Victor went to sleep; it was a particular and most unusual kind of sleep, one that Puységur had not witnessed or attended to before. The trance-state was what later times would call 'hypnosis'. The medical historian, Henri Ellenberger, writing two centuries later, in our times, describes the event:

"There were no convulsions, no disorderly movements, as was the case with other patients: rather, he fell into a strange kind of sleep in which he seemed to be more awake and aware than in his normal waking state. He spoke aloud, answered questions, and displayed a far brighter mind than in his normal condition. The Marquis, singing inaudibly to himself, noticed that the young man would sing the same songs aloud. Victor had no memory of the crisis once it had passed. Intrigued, Puységur produced this type of crisis again in Victor and tried it successfully on several other subjects." [3]

Adam Crabtree, also a modern historian of these events, provides a quotation translated from Puységur's 1784 account:

"He [Victor Race] spoke, occupying himself out loud with his affairs. When I realized that his ideas might affect him disagreeably, I stopped them and tried to inspire more pleasant ones. He then became calm, imagining himself shooting a prize, dancing at a party, etc. I nourished these ideas in him and in this way I made him move around a lot in his chair, as if dancing to a tune; while mentally singing it, I made him repeat it out loud. In this way I caused the sick man from that day on to sweat profusely. After one hour of crises, I calmed him and left the room. . . . He slept all night through. The next day, no longer remembering my visit of the evening before, he told me how much better he felt." [4].

Mesmer himself had been aware of the fact that certain patients appeared to sleep: these were called 'somnambulists', but it was left to Puységur and other observers to notice that this sleep was of an unusual kind, for the mind of the subject appeared to be awake while the subject physically slept. Puységur noticed that young Victor seemed able to hear only Puységur's voice. After Puységur told Victor that he, Victor, should place a magnetized cord on his chest, and that he would be thereby healed, Victor was told to awake and be healed. He did; and he was.

Massachusetts, U. S. A., 1876: Professor Fowler Examines Mr. Phillips' Skull

On September 16, 1876, Mr. E. W. Phillips paid a visit to 'Professor' Fowler. The self-titled 'Professor' was touring Massachusetts, offering readings, away from his base at 'The Phrenological Cabinet', this a shop at 135 Nassau Street in lower Manhattan. The shop was said to be the second most-visited site by tourists in New York City, one whose popularity was exceeded only by Mr. Barnum's uptown 'Museum of Oddities'. [5] Fowler's publications and paraphernalia, chiefly casts of skulls with descriptions of the psychological correlates expressed by the shape of the skull, sold well. There were also displays of skulls that could be viewed at no charge and instructional guides as to how to practice phrenology. Best of all, a staff was available to read for a few the skull of visitors.

As was true of hypnotism, the skills of phrenology were easily learned. Nearly anyone could become a practitioner by following directions on how to feel a skull. To add to the possible frisson of research, Fowler and others, mostly relatives, published pamphlets showing which aspects of the head and body to examine 'in privacy'.

Although Phillips' motivations for requesting a reading are unknown, the outcome of the visit is certain, for I hold it in my hand. It is a copy of Professor Fowler's book, purchased by Phillips from Professor Fowler. It contains Fowler's measurements of Phillips' skull, along with indications from these measurements of Phillips' talents, skills, and predicted romantic successes. [6]

Douglas Keith Candland

THE

PRACTICAL PHRENOLOGIST;

AND

RECORDER AND DELINEATOR

OF THE

CHARACTER AND TALENTS

OF

E. W. Phillips.

As marked by

Prof. O. S. Fowler

A COMPENDIUM

OF

PHRENO–ORGANIC SCIENCE.

BY

O. S. FOWLER,

PRACTICAL PHRENOLOGIST, LECTURER, FORMER EDITOR OF "AMERICAN PHRENOLOGICAL JOURNAL"
AND AUTHOR OF "FOWLER ON PHRENOLOGY," "FOWLER ON PHYSIOLOGY," "SELF-CULTURE"
"MEMORY," "RELIGION," "MATRIMONY," "HEREDITARY DESCENT," "LOVE AND
PARENTAGE," "MATERNITY," "AMATIVENESS," "SELF INSTRUCTOR," "HOME
FOR ALL," "ANSWER TO HAMILTON," "VENDEX," ETC., ETC., ETC.

BOSTON :

O. S. FOWLER, 514 TREMONT STREET.

Sept 16th 1876.

Figure 2.1 shows the title page of the book that Mr. Phillips
purchased from Professor Fowler..

24

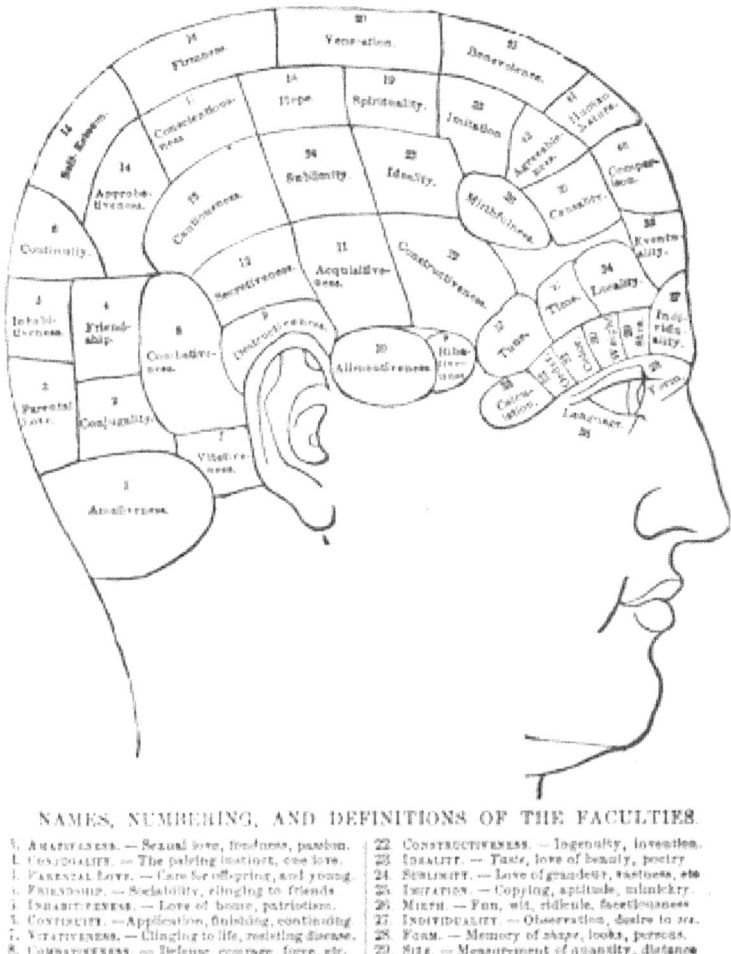

NAMES, NUMBERING, AND DEFINITIONS OF THE FACULTIES.

1. AMATIVENESS. — Sexual love, fondness, passion.
2. CONJUGALITY. — The pairing instinct, one love.
3. PARENTAL LOVE. — Care for offspring, and young.
4. FRIENDSHIP. — Sociability, clinging to friends.
5. INHABITIVENESS. — Love of home, patriotism.
6. CONTINUITY. — Application, finishing, continuing.
7. VITATIVENESS. — Clinging to life, resisting disease.
8. COMBATIVENESS. — Defense, courage, force, etc.
9. DESTRUCTIVENESS. — Executiveness, severity.
10. ALIMENTIVENESS. — Appetite, relish, greediness.
11. ACQUISITIVENESS. — Frugality, saving, industry.
12. SECRETIVENESS. — Self-control, policy, art, tact.
13. CAUTIOUSNESS. — Guardedness, safety, prudence.
14. APPROBATIVENESS. — Pride of character, honor.
15. SELF-ESTEEM. — Self-respect, dignity, austerity.
16. FIRMNESS. — Stability, perseverance, willfulness.
17. CONSCIENTIOUSNESS. — Duty, right, truth, justice.
18. HOPE. — Expectation, anticipation, enterprise.
19. SPIRITUALITY. — Intuition, prescience, faith.
20. VENERATION. — Worship, adoration, obedience.
21. BENEVOLENCE — Sympathy, kindness, goodness.
22. CONSTRUCTIVENESS. — Ingenuity, invention.
23. IDEALITY. — Taste, love of beauty, poetry.
24. SUBLIMITY. — Love of grandeur, vastness, etc.
25. IMITATION. — Copying, aptitude, mimicry.
26. MIRTH. — Fun, wit, ridicule, facetiousness.
27. INDIVIDUALITY. — Observation, desire to see.
28. FORM. — Memory of shape, looks, persons.
29. SIZE. — Measurement of quantity, distance.
30. WEIGHT. — Control of motion, balancing.
31. COLOR. — Discernment and love of colors.
32. ORDER. — Method, system, doing by rule.
33. CALCULATION. — Mental arithmetic, reckoning
34. LOCALITY. — Memory of place, position, etc.
35. EVENTUALITY. — Memory of facts, events, etc.
36. TIME. — Telling when, time of day, dates, etc
37. TUNE. — Musical love, ecstacy, and talent
38. LANGUAGE — Expression by words, acts, etc
39. CAUSALITY — Planning, thinking, reason, sense
40. COMPARISON. — Analysis, inferring, critic
41. HUMAN NATURE. — Perception of character
42. SUAVITY. — Pleasantness, blandness, blarney

Figure 2.2, from later in the same booklet by Fowler presented to Mr. Phillips, shows the areas of the skull and the talents and skills to be found represented there.

Douglas Keith Candland

What Mr. Phillips learned about his abilities and proclivities from the reading by Professor Fowler of his skull is shown in Figures 2.2 and 2.3. What Phillips learned about himself seems promising: Under 'Business Adaptations' we, as did Phillips, see that

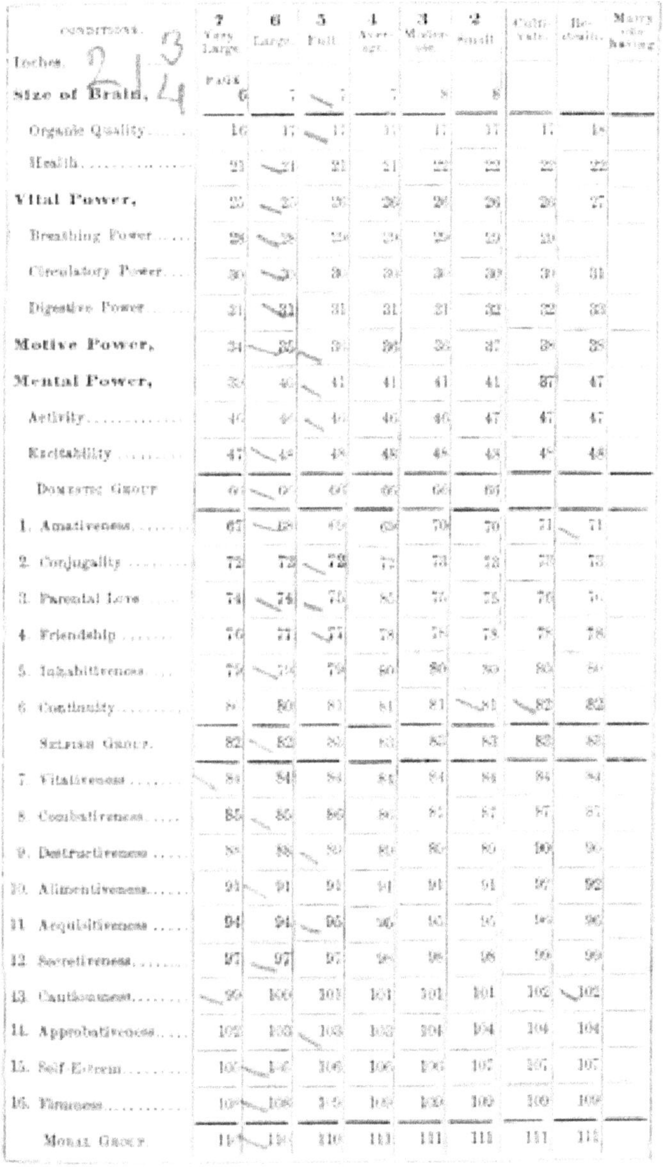

CONDITIONS.	7 Very Large	6 Large	5 Full	4 Average	3 Moderate	2 Small	Cultivate.	Restrain.	Marry one having
Inches. 21 3/4									
Size of Brain,	PAGE 6	7	7	7	8	8			
Organic Quality	16	17	17	17	17	17	17	18	
Health	21	21	21	21	22	22	22	22	
Vital Power,	25	25	26	26	26	26	26	27	
Breathing Power	28	28	28	29	29	29	29		
Circulatory Power	30	30	30	30	30	30	30	31	
Digestive Power	31	31	31	31	31	32	32	33	
Motive Power,	34	35	35	36	36	37	38	38	
Mental Power,	39	40	41	41	41	41	87	47	
Activity	46	46	46	46	46	47	47	47	
Excitability	47	48	48	48	48	48	48	48	
DOMESTIC GROUP	66	66	66	66	66	66			
1. Amativeness	67	68	69	69	70	70	71	71	
2. Conjugality	72	72	72	72	73	73	73	73	
3. Parental Love	74	74	75	75	76	75	76	76	
4. Friendship	76	77	77	78	78	78	78	78	
5. Inhabitiveness	79	79	79	80	80	80	80	80	
6. Continuity	80	80	81	81	81	81	82	82	
SELFISH GROUP.	82	82	82	83	82	83	83	83	
7. Vitativeness	84	84	84	84	84	84	84	84	
8. Combativeness	85	85	86	86	87	87	87	87	
9. Destructiveness	88	88	89	89	89	89	90	90	
10. Alimentiveness	91	91	91	91	91	91	92	92	
11. Acquisitiveness	94	94	95	96	95	95	96	96	
12. Secretiveness	97	97	97	98	98	98	99	99	
13. Cautiousness	99	100	101	101	101	101	102	102	
14. Approbativeness	102	103	103	103	104	104	104	104	
15. Self-Esteem	105	106	106	106	106	107	107	107	
16. Firmness	108	108	109	109	109	109	109	109	
MORAL GROUP	110	110	110	111	111	111	111	112	

Figure 2.3 The First part of the reading of Mr. Phillips' skull. Continued in 2.4.

Phillips's future is best served by his becoming a baker, boss workman, or farmer, and a little less successfully by his becoming a carpenter. Measurement of the various localities of the skull suggest that Phillips is both 'vital' and 'cautious' as well as 'conscientious', 'benevolent', and 'constructive' (these on a page not shown). He is urged to show restraint in 'amativeness' (demonstrating physical love), seemingly contradictorily to the assessments of natural cautiousness and benevolence. Even family planning is included, as he is urged to marry one who has 'hope', 'perceptiveness', 'individuality', and 'literary faculties'.

CONDITIONS.	7 Very Large	6 Large	5 Full	4 Aver age.	3 Moder ate.	2 Small.	Light rate	Ro. stras	Marry one having
17. Conscientious	112	113	113	113	114	114	114	114	
18. Hope	115	115	115	116	116	117	117	115	
19. Spirituality	117	118	118	118	118	118	119	118	
20. Veneration	119	120	121	121	121	121	121	121	
21. Benevolence	122	123	123	123	123	123	123	124	
SELF PERFECTIVES.	129	124	124	124	124	126	125	125	
22. Constructiveness	125	125	126	126	126	126	126	127	
23. Ideality	127	127	128	128	128	128	128	128	
24. Sublimity	128	129	129	129	130	130	130	130	
25. Imitation	130	131	132	132	132	132	132	132	
26. Mirthfulness	133	133	134	134	134	134	134	134	
INTELLECTUALS.	135	135	135	135	135	135	135	135	
PERCEPTIVES.	136	136	136	137	137	137	137	136	
27. Individuality	137	138	138	138	138	138	138	138	
28. Form	138	139	140	140	140	140	140		
29. Size	140	141	141	141	141	141	141	142	
30. Weight	142	142	142	142	143	143	143	143	
31. Color	143	143	144	144	144	144	144	144	
32. Order	144	145	145	145	145	145	146	146	
33. Calculation	146	147	147	147	147	147	147	147	
34. Locality	147	147	148	148	148	148	148	148	
LITERARY FACULTIES.	148	148	149	149	149	149	149	149	
35. Eventuality	149	149	150	150	150	151	151	151	
36. Time	151	151	152	152	152	152	152	152	
37. Tune	152	152	152	152	153	153	153	153	
38. Language	153	154	154	154	154	155	155	155	
REFLECTIVES.	155	156	156	156	156	156	156	156	
39. Causality	157	157	157	158	158	158	158	158	
40. Comparison	159	159	159	160	160	160	160	160	
41. Human Nature	160	161	161	161	161	161	161	162	
42. Agreeableness	162	162	162	162	162	162	162	162	

Figure 2.4: Continuation of 2.3 showing Professor Fowler's readings of Mr. Phillips skull.

Although some phrenologists' charts include 'two persons' measurements, presumably one of each of a couple seeking advice on their degree of compatibility, Mr. Phillips' record remains unadorned by that of a companion, and we do not know what the future held for him either in profession or amativeness. We do not know whether the forecast itself influenced his behavior and aspirations. We do not know whether the judgments were reliable, in the sense that they could be found by independent observers, or valid, in the sense that the judgments represent meaningful categories of behavior.

New York, U. S. A., 1889: Phrenology Expands

To some adherents, the shape of the skull was only one among many aspects of the body that revealed information about a person's temperament, abilities, and talents. For the practicing phrenologist, a trained look at the facial features, hair and eye color, and shape of the skull was revealing. Consider the faces of the three young men pictured in Figure 2.5. The phrenologists, in their book *Heads and Faces, How to Study Them,* told us that one man will succeed at business, one at book-learning, and one at working with mechanisms. You may judge which is best suited for which occupation.

Figure 2.5: Three young men. Their faces predict the professions in which they will excel.

The 'correct' answer is as follows:

The man on the left should be destined for a life of working with mechanisms.
"He is assessed as capable of enduring the fatigue and hardship which may be imposed on the man who is a builder or a mechanist. The fullness of the brows indicates practical judgment; the head broadens out, backwards from the corner of the eye, and upward towards the temples where Constructiveness, Acquisitiveness, Ideality, and Sublimity are located, giving him the faculty for invention, mechanical ingenuity, the ability to construct anything from a watch to a locomotive.

The young man in the middle should consider business.

". . . a bright face and a well developed head. His large Language indicated by the fullness of the eye and the width and prominence of the brow qualify him to take in the particulars pertaining to business affairs; to recognize details, and to describe whatever he knows about a subject. The forehead shows planning ability."

The man on the right should consider books.

". . .the young man on the right, will be happiest and most successful for a life of the mind because of the distance between chin and eyebrows, compared to the width of the eyebrows. [7] The tallness of the head indicates spirituality, reverence, integrity, imagination — all well-known characteristics of teachers, scholars, and writers, to be sure."

Like specifications are offered for the lawyer, for whom language is most important; the clergyman, for whom 'subordinate properties' are essential; and the physician, for whom 'combativeness' is regarded as necessary to perform surgery. The list includes a man to care for horses who should have large 'Philoprogenitiveness', and advice is given regarding boys with large brains and small bodies. These fellows should not be given a college education, it is suggested, but should be on a farm, where the body can be developed.

That acceptance of phrenology was a characteristic of the times in Europe and America is shown by the example of a believer — Walter (Walt) Whitman, then a reporter for the newspaper, the *Brooklyn Eagle*, but eventually one of America's most admired poets. He wrote favorably of the Phrenological Cabinet in his newspaper columns. More than once he had his own skull read at the Cabinet. Coincidentally, it was the Fowler family, through their publishing empire, that published Whitman's *Leaves of Grass*. The original publisher bowed out after he read the final manuscript and determined the work to be pornographic. (Chapter 5)

New Jersey, U. S. A., 1868 to New York, U. S. A., 1980: Darwinism Meets Human Progress

Accommodation of evolution in schools was not easy. Most American colleges and universities in the late nineteenth century were founded and controlled by religious groups holding to the ideal of combining education and evangelicism. There was, to be sure, an emerging set of state-supported land-grant colleges, state legislatures having thought it wise to create institutions that assisted residents in gaining practical skills, such as farming and machining. The idea of humankind and animals's sharing their modes of origin encountered resistance in those as well, for elected representatives can be as theocratic as clergy. There were, however, some brave, pioneering institutions.

In 1868, Princeton University (then the College of New Jersey) appointed the Scottish clergyman, James McCosh, president in the hope that he would advance the little college by removing the intellectual stodginess that the trustees believed to be rampant among the faculty. To overcome it, they chose a person who, in time, married Darwinian ideas of evolution to the religion of predestination. He did this by introducing paleontology, psychology, and evolution into the 'stodgy' curriculum.

McCosh was converted to the idea that God and Darwinism could co-exist. As he informed his classes, 'if

God is all-powerful, it may please God to establish natural selection and variability: 'The doctrine of evolution does not undermine the argument from Final Causes, but rather strengthens it by furnishing new illustrations of the wisdom and goodness of God.' [8]

The groundwork for this fusion had been planned, planted and cultivated by the influential Professor E. D. Cope, who accepted and taught the Lamarckian idea of the 'inheritance of acquired characteristics'. With acceptance of this idea came the notion that body and mental parts not used would eventually disappear. Those used, or 'adaptive', would be 'selected for'. Inherent in this sort, he argued, was a notion of teleology, the belief that there is a purpose or final state toward which the universe and all within it are moving.

To digress in time for a moment: Professor Cope would acquire another distinction, this one not granted to any other human being. The story is best told by its describer, Michael Sims:

"The first member of a new species to be described becomes its 'type specimen' the representative by which the species' unique features are delineated. . . . By the standards now in use, because Linneaus described the human species without designating a type specimen, the scientific name he chose could have been challenged at any time until a lectotype [an elected type specimen] could be appointed. That didn't happen until 1994. . . . Although he [Cope] has remained legendary among the eccentrics who populated the field of paleontology in the nineteenth century, his deathbed wish was the most eccentric of all. . . Cope wanted his skeleton to become the type specimen for Homo sapiens. Unfortunately, after Cope's death in 1897 his skeleton was deemed unsatisfactory to fill the noble role for which he had volunteered. Apparently his bones exhibited the early stages of syphilis . . ."

In 1994, ninety-seven years after his death, an amused scientific panel approved the nomination, and Edward Drinker Cope officially became the elected type specimen of Homo sapiens. At the same time, his place of birth became the representative human type locality — Philadelphia." [9]

Returning us to McCosh's Princeton: McCosh's hired "new men" who accepted the view that genetic material was formed at conception; that it was unchangeable by environment except through truly dramatic forces such as radiation, but surely education, socialization, acculturation, and the like had no effect on the gene itself. There was no evolution by means of acquired or learned talents or characteristics being passed along genetically. Natural selection worked slowly because it worked only once in each lifetime, at the time of the creation of each new collection of genes. [10]

Among those McCosh selected for the new faculty was the then young Henry Fairfield Osborn. Through his later stewardship of the American Museum of Natural History in New York City (1890-1935), he influenced generations of adults' and children's understanding of humankind and its origins. Osborn's influence through the public pulpit provided by the displays at the museum was marked by his notion of a ladder of mental and physical ability. The continuum arranged by Osborn included species of animals and races of human peoples, leaving the clear impression that human progress had resulted in human beings' perching uncontested atop the ladder. Fossils and remains were placed in a like manner by the arranging of dioramas that displayed extinct peoples and their cultures. These constructions, shown in Figure 2.6, made visual statements about the development of human cultures, showing how human races and species evolved. The dioramas were removed in the 1980s.

The dark Neanderthals were displayed as limited to small-time hunting and living more or less in a Neolithic slum. The lighter-skinned Cro-Magnon, on the contrary, lived in a veritable suburban paradise of civility, doing artwork in the evenings by the light of the lamps they had built after they discovered how to tame fire. The even lighter-skinned Saxons, dressed in animal skins (as do some people today), had a leader, and looked ready to read Nietzsche for self-improvement.

Figure 2.6: Murals from the American Museum of Natural History from the pictures of the American Museum of Natural History by Charles R. Knight. Description is in Freeland and Adams. Murals are copies from Rainger, 1991.

Osborn's other interests included 'human eugenics', this being the movement to improve the human species by regulating reproduction. The Museum hosted the Second and Third International Congress of the Eugenics Society, a group headed at one point by a son of Charles Darwin. Osborn also served on boards directing research on human genetics. A committed conservator, he used his influence and authority to protect the California redwoods from logging. Osborn and his colleagues believed in presenting 'living displays'. Human displays in zoos and museums were not unusual at this time, and chief among these were the Eskimos whom Admiral Peary had 'recruited' from Lapland. (Chapter 3)

Germany, 1907 and the United States, 1920: At the Asylum

In 1907, twenty-eight year old Ernst Kretschmer, a newly graduated physician, was the director of what was then called an asylum. He had already begun the observations that would be published in his now classic book Physique and Character, in which he investigates the relationship between body-types and psychotic behavior. Kretschmer writes:

"In the mind of the man-in-the-street, the devil is usually lean and has a thin beard growing on a narrow chin, while the fat devil has a strain of good-natured stupidity. The intriguer has a hunch-back and a slight cough. The old witch shows us a withered hawk-like face. Where there is brightness and jollity we see the fat knight Falstaff red nosed and with a shining pate. The peasant woman with a sound knowledge of human nature is undersized, tubby, and stands with her arms akimbo. Saints look abnormally lanky, long-limbed, of penetrating vision, pale, and godly." [11]

Kretschmer's work on body-typing varies from phrenology in two ways: it is concerned with the temperaments of those classified as 'mad' and, second, it makes a serious statistical attempt to measure these differences. From approximately 1905 until 1915, Kretschmer and his staff rated inmates' physiques by using a complex scale described initially as "Constitutional Scheme" and in later years as 'Constitutional Psychology'. (Chapter 5).

Being director of an asylum, Kretschmer had ample occasion to observe various forms of insanity. He also had the authority to require the inhabitants be inspected, measured, and photographed. His descriptions of these people are perspicacious and compassionate. The manic-depressive, according to Kretschmer, has 'circular insanity' because the excitement of the mania is followed in the cycle by depression. Kretschmer's observation was that manic-depressives tended toward physical 'compactness', although by today's standards of leanness the photographic examples look lumpy. He described them using the Greek word, 'pyknic.' Schizophrenics, he observed, tended toward fragility. He describe them using the word 'asthenic' (meaning 'without strength') perhaps referring to the observation that schizophrenics loose both motivation and the affective emotions during their illness. He classified a third type as 'athletic'.

Kretschmer and his assistants identified each person photographed as pyknic or asthenic, thereby separating the manic-depressive from the paranoid schizophrenic, as shown for example in Figure 2.6. Good prediction, of course, works both ways: one ought to be able to predict the body type from the diagnosis, and the diagnosis from the body-type. Alas, Kretschmer and his staff knew the diagnosis before examining the photographs. This seeming lack of elementary attention to experimental control is characteristic of the nascent social science of that day. As we shall see, however, such experimental contamination extends into our own times.

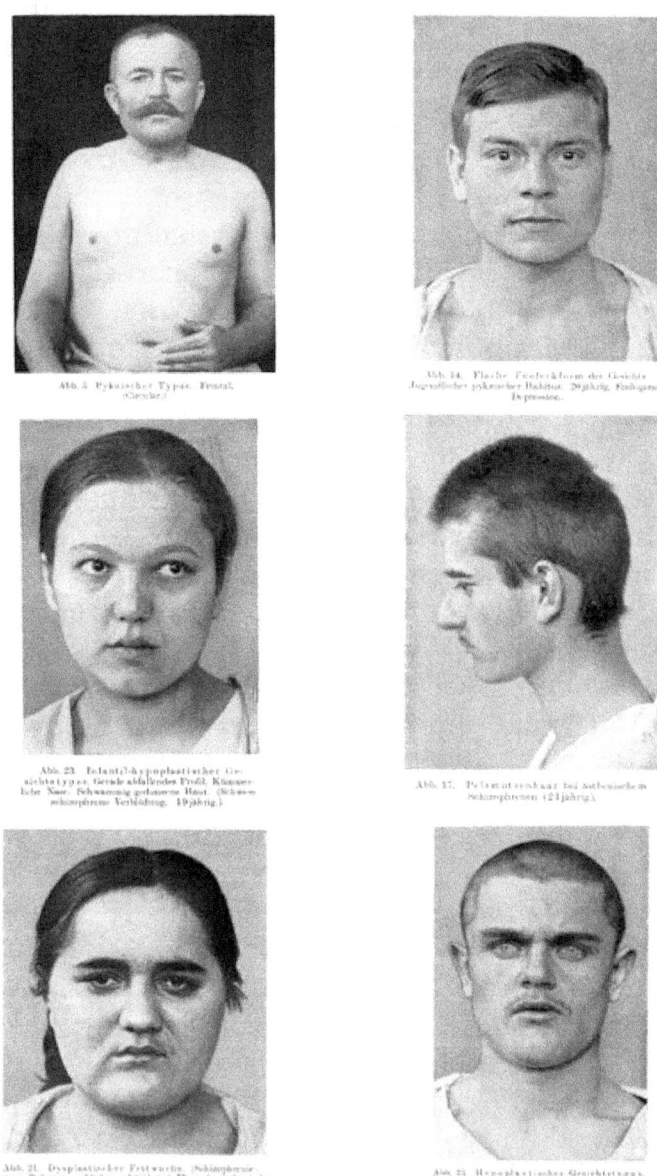

Figure 2.7: Faces and Body-types, personality, and psychoses, from Kretschmer, 1925.

In the 1920s, the American William Sheldon [1898-1977] began a large-scale development of a system based on Kretschmer's ideas. He took photographs of men and women whose body-types were then measured by assigning numerical scores to aspects of the body. Sheldon's work began at the Universities of Chicago and Texas (which awarded him both the Ph.D. and M.D.) and continued at Wisconsin, Harvard, Oregon, and Berkeley. His goal was to relate physique to personality, temperament, and the prediction of criminal and antisocial behavior. He received approval from colleges and universities to photograph enrolling male undergraduates, and some females, willing or required, to have their bodies photographed and classified.

Sheldon's first work measured and classified the body-types of residents of a home for young males regarded as delinquents. His expressed interest was to describe their temperament upon entering the home so that they could be "housed appropriately". Sheldon and his colleagues developed a system by which the human physique was divided into one of three main types: endomorphy, mesomorphy, and ectomorphy. Each was categorized by a seven-point scale with four being neutral. The system yields 343 possibilities, such as 5-6-5, 2-7-3, and the like. A 'perfect' mesomorph, for example, is a 4-4-4.

An example is that of an 18-year-old 'extreme endomorph', characterized as a 1-1-7. Sheldon writes, hopefully with tongue in cheek, but in a manner characteristic of his sometimes flipant interpretations:

"The 1 1 7 with his extreme predominance of surface over mass, seems caught in a predicament of biological overexposure, and for such an organism the ordinary

41

circumstances of social life may in fact amount to chronic overstimulation. Hebrephrenic psychopathy may be one natural response to (way out of) such a situation. The 1 1 7 is more common in the mental hospitals than in the general population, and his diagnosis is usually hebeprehnic schizophrenic. But also he is encountered more frequently on college campuses than in the general population, and there the diagnosis is sometimes Phi Beta Kappa. This occasional coincidence of sensitive brilliance and hebephrenic jettisoning is one of the things that make mental tests difficult to interpret, particularly when the somatotype is not known." [12]

Sheldon's initial assessments made much of presumed gender differences. The young men were rated on their degree of 'femininity': ability at boxing was admired, but talent at dancing was scored a feminine trait. When examining the work of the 1930s, the contemporary reader is struck by the importance given to what Sheldon called the DAMP RAT syndrome, this an acrostic for a set of presumably related, diagnostic temperamental traits; in this case Dilettante, Arty, Monotophobic, Perverse, Restive, Affected, Theatrical. (13)

Some males, we are informed, are maternal, and "does his mothering with no trace of any DAMP RAT characteristics. Most of the male homosexuals in this series [i.e., the group of men being studied] are DAMP RATS, although not all DAMP RATS are homosexuals. There are many men who live in the 'arty perverse' DAMP RAT pattern with never a suggestion of homosexuality. [14] (Chapter 6)

Italy, 1876: Predicting the Criminal

When searching for distinguishing characteristics of potential criminals, Cesare Lombroso, himself born into a Jewish family, readily separated Jews as a distinct racial type given to criminality. His first book, *The Human Delinquent*, published in 1876, established him as the leader of the view that regarded criminality as largely inherited, rather like epilepsy. Criminals, he demonstrated to applause, are a lower order of humanity, just as Negroes, he thought, were an order of humanity somewhat below Caucasians. Lombroso looked into the brain of dead criminals to find the location of criminality. He found a 'distinct depression' which he named the 'median occipital fossa', this being a protuberance of the brain not seen clearly since [15]. That Lombroso did not think to examine the brains of noncriminals as a control helps to give the descriptions of his methods, at least those of his earliest work, the charm of the observant naturalist.

Lombroso recalls the development of his ideas:

"The first idea came to me in 1864, when, as an army doctor, I beguiled my ample leisure with a series of studies on the Italian soldier. From the very beginning I was struck by a characteristic that distinguished the honest soldier from his vicious comrade: the extent to which the latter was tattooed and the indecency of the designs that covered his body. This idea, however, bore no fruit.

The second inspiration came to me on one occasion, amid the laughter of my colleagues, I sought to base the study of psychiatry on experimental methods.

When in '66, [1866] fresh from the atmosphere of clinical experiment, I had begun to study psychiatry, . . . I applied to the clinical examination of the skull, with measurements and weights, by means of the esthesiometer and craniometer. Reassured by the result of these first steps, I sought to apply this method to the study of criminals." [16]

Tellingly, Lombroso picks up on the idea of variability, concluding, "I was anxious to apply the experimental method to the study of diversity, rather than the analogy, between lunatics, criminals and normal individuals . . . This method proved useless for determining the difference between criminals and lunatics". [17] He later therefore assumed, using inexorably bad logic, that criminals and lunatics are alike. However, Lombroso did derive a theory for distinguishing the criminal from the normal by combined knowledge of three famous criminals, with which he had various degrees of familiarity.

Lombroso was among those who performed an autopsy on the "famous brigand" Vilella. In life, Lombroso tells us, Vilella was so agile as to carry a sheep on his shoulders while scaling mountains, but capable of murder and rape. In the brain of dead Vilella, Lombroso notices the unusual size of the 'median occipital fossa'. A second criminal, named Verzeni, reinforced Lombroso's idea that criminal behavior was primitive; that the criminal is a sub-human or not-quite-yet-human. Verzeny, by showing "cannibalistic instincts of 'primitive anthropophasigists' (the love of eating one's own species) and the ferocity of beasts of prey," demonstrated how heredity determined the ladder of socialization. A third criminal studied by Lombroso, was reported as a young man to be normal in every way except for epileptic seizures. One day, for no apparent reason, he killed eight officers, then slept soundly for twelve hours, and awakened without memory of the events.

Based on these three persons and their acts, Lombroso forms a theory of the causes of criminality: these being an uncommon brain structure, subhuman genes, and seizures of the brain for which the sufferer has no memory. The American edition of Lombroso's pamphlet arguing, from what he thought was a Darwinian point of view, that criminal behavior was genetic, was eventually published by Professor Fowler, the phrenologist.

.

When Lombroso published of the fifth edition of his work, *Human Delinquents,* he had added measurements of many social and economic variables, such as the nature of the economy, international trade, and the climate of nations. In his last work, this published in 1911 in English in order to reach the large market, wherein his ideas were now finding welcome, his observations lead him to offer the hypothesis of their being three kinds of criminals: the born, the insane and mentally deficient (these bundled together), and 'criminaloids,' The last were people not disposed to crime because of their genes, but capable, because of their environment, of criminal behavior under special circumstances. Regardless of one's genes, one might steal a loaf of bread if one's children were hungry. (Chapter 7)

The U. S. A., 1890 to the present: Three Generations of Imbeciles are Enough

Following the work of a certain Dr. Sharp, who vasectomized young male inmates for the state of Indiana in 1899 with the stated hope of eliminating 'excessive' masturbation', sterilization laws were passed by the states of Connecticut, Washington, California [1909], New Jersey, Iowa [1911] Nevada, New York [1912] North Dakota, Michigan, Kansas and Wisconsin [1913] Nebraska [1915] and Oregon, South Dakota, and New Hampshire [1917]. [18]

These laws had been challenged and often rejected at the upper levels of state courts. In 1927, the United States Supreme Court announced its decision in the case of Buck v Bell, which attempted to challenge laws at the highest level. The case concerned whether the state of Virginia had a legitimate and constitutional interest in sterilizing, without her knowledge, a young woman then residing in a state-run facility. Justice Oliver Wendell Holmes wrote a decisive decision, in which he said that:

"It is better for all the world, [he wrote] if instead of waiting for their imbecility, society can prevent those who are manifestly unfit from continuing their kind. The principle that sustains compulsory vaccinations is broad enough to cover cutting the Fallopian tubes . . . Three generations of imbeciles are enough". [19] (Chapter 7)

We Go to the Fair

Having examined these small finds, we must select a first site for an extended excavation and set forth. Some of our finds suggest that we take a trip to the Chicago World's Fair, for many whom we have seen either appeared there, or used methods first publically highlighted there. Let us pretend that it is a clear April day in 1893, and the six-month run of the fair has just begun. The opening date missed the famous four-hundredth anniversary of Columbus's discovering the

Americas for Europeans by only a few months. From the descriptions, histories, and pictures available, we may find many remnants of what people then believed, always a starting place to finding mental fossils.

Chapter 3

The Congress of Ideas

"She said, 'I suppose a museum is a celebration of death. Dead people's lives, the objects they made, the things they thought important, their clothes, their houses, their daily comforts, their art.'

"No. [he replies] a museum is about life. It's about the individual life, how it was lived. It's about the corporate life of the times, men and women organizing their societies. It's about the continuing life of the species Homo sapiens. No one with any human curiosity can dislike a museum." P. D. James, The Murder Room, p. 63.

For us, prospective, pioneering, and inexperienced archeopsychologists, examination of fairs offers an unmatchable opportunity for an exploratory dig. When we unearth what we know about the fair from the then contemporary sources, the fair offers us a rich source of fossils from which to begin our study of the archeology of the mind. For one thing, the world fairs offers at a single location a brief window into whatever people of the time thought to be the most advanced states of knowledge, culture, and human achievement. For another, the ways in which exhibits are organized and described says much about the mentality of the times. Thirdly, the exhibits are usuall well-documented by minutes of meetings in which we find out what exhibits were chosen or rejected and why. We mental-fossil-hunters are interested in what –

themes the fair offered, the fair's perception of its educational mission, and evidences of the sort of material displayed with awe and pride. We select this fair in particular because it connects together our previously suggested finds (described in Chapter 2) with the ideas we hope to analyze.

The theme of the 1893 Chicago Fair was 'Human Progress', this theme alerting us to how people, at least people of the Midwestern United States, understood their place in time. We unearth evidence that nineteenth century ideas, those of the evolutionists Darwin, Wallace, Spencer, and the statistician Galton, along with the already-mentioned phrenologist, Fowler, stimulated an interest in the differences and variations among people and peoples, leading to a categorization of people in terms of their 'cultures' and 'races.' As we stroll the fair grounds, we shall be struck by the pervasiveness of 'differences and variation' as attached to the idea of 'human progress', for the displays of living human beings are grouped according to their cultural evolution. The Chicago fair is also particularly exciting because it also offered the first public display of achievements for three new fields of inquiry into the human mind - anthropology, neurology, and psychology.

"Differences and Variation" could have been the fair's theme rather than the one selected, 'Human Progress", for Galton's inventions of statistical measures and the application of these to human variations in skills and abilities lead directly to a false sense of the validity when measuring human capacity, a sense that can be linked with the practice of eugenics (Chapter 8). The intellectual air was alive with the idea of mental and physical evolution, this from Darwin, Wallace, and, especially

Spencer's interpretation of the 'survival of the fittest' as the politics of social Darwinism. Darwin and Wallace's discovery depended on diversity of structure, mentalities, and although they did not know it, depended on genetic diversity. These, the basic issues of 20th century Europe and America, where there for all to see who had the prescience to think beyond the dazzling displays of inventions and the displays of peoples arranged by race and culture.

The Old and the New

Chicago wished to advertise its return from ashes, for much of the downtown and many homes had burned only twenty or so years before, leaving half the city in a depression of a sort not to be re-experienced until the New York World's Baseball Series of 1984. Through the fair, Chicago wished to draw attention to its rebirth by joining the other great cities that had sponsored world fairs: Paris twice, most recently in 1889; London in 1885, featuring its great display in the Crystal Palace; Philadelphia, for the nation's 1876 centennial. St. Louis's celebration was twelve years in the future: as was Buffalo, which would be the locale of a presidential assassination; New York yet to come, twice, San Francisco, also twice, and Chicago, once again in 1933. [1]

Chicago offered itself as the celebrant of the four-hundredth anniversary of Columbus's discovery of the new world, although it is difficult to think of a direct connection between Columbus and Chicago. Nonetheless, the Duke of Veragua and his family sat on the speakers' platform on opening day, May 1, 1893, as he, and they, were reputed to be direct descendants of Christopher Columbus. The grounds were dazzling, as we can recreate from Figure 3.1 and 3.2. Those who traveled to Chicago were treated to the sight of buildings ringed with the new electric light-bulbs, a sight especially stunning at night because of the reflection from waterways and lakes throughout the fairgrounds, this a shrewd way to use the area's swampiness to effect. Not only did electricity create daylight from darkness in a way previously unknown, but also various displays used electricity to produce movement. The large motors and engines, especially those offered by German industry, showed the awesome, unbelievable, power of this curious, invisible force.

Who would have imagined that something which itself could not be seen - something whose source was invisible — could offer such power? In Chapter 4, we shall offer a comparison between electricity (as an unseeable but powerful force) and spiritism (a belief in unseeable alternative worlds and powers) and suggest that these form the origins of experimental psychology through the efforts of psychophysics.

Douglas Keith Candland

Figure 3.1: The Chicago Fair of 1893. Through the 'unseen power' of electricity night became day. Another 'unseen power', resulting in photography, graced the displays of the new sciences represented, anthropology, neurology, and psychology. Another unseen power, telegraphy, was also a marvel.

At the Fair

The Chicago Exposition was successful in showing the marvels of the world, both living and manufactured. If the planners' goal was to exhibit the extent of humankind and its achievements, the goal was surely both met and exceeded. Never before had such abundances of discovery been set out in a single site; never had so many men and women of achievement in art, manufacturing, science, religion, and education appeared together in a short span of time to show and explain their work and ideas.

Peace was extolled. The Exposition's president announced at the opening-day festivities that the Exposition came at a time when "A new leader has taken command. The name of this leader is Peace." Alas for veracity, the mayor of Chicago had been assassinated the night before by an unhappy worker, while the United States and Spain would be at war within five years, Germany and Great Britain in eleven, and much of the world, shortly thereafter.

The Exposition Grounds contained the Great Hall, which displayed art, crafts, and manufacturing, items displaying the pride of the countries participating. The German display of Krupp weaponry, especially the artillery, was remarked upon with awe, while the British disappointed by offering only a house in which British-made furniture was displayed, this to the chagrin of Chicagoans who understood the unimaginative showing to represent a lack of respect. Documents of the time make evident the praise offered the German-speaking nations, whether for their weapons, colonial power, or cakes, while the British offerings were perceived as too little, too late, and too old, as if they were claiming reverence for their past and not the expectations of the progress expected from the future.

The construction of a Women's Hall represented more than the symbol that men and women differed in their needs by also expressing a newly-found power among women. The plan of building a women's center was put forth by Mrs. Potter Palmer. No doubt, the Palmers' generosity and social standing, no small part of which had been used to rebuild Chicago, gave her views a certain clarity to the fair committee. Her request for a separate building for and honoring women was seen as an appropriate one, for the Congress of the United States, which provided part of the monies needed for the fair, had agreed to the sensibility of a special building featuring and accommodating women and women's work. If women were not yet believed to be the equal of men, thereby qualifying themselves for voting, their achievements were nonetheless judged worthy of a certain featuring, if a separate one.

European and American art were shown of such diversity and number that, in a day in which reproduction of images was poor, it was an assemblage of visual images previously unknown. But one does not go to a fair only to look at European paintings, displays of the proud states, examples of manufacturing, displays of education and homecraft, or even electric lights. Connected to the area of the Exposition Buildings, but separated from it by design, was the Midway-Plaisance, a strip that extended almost one mile from the elegant grounds and Renaissance buildings. Here entertainment and refreshments were provided in abundance.

The activities and attractions of the Fair turn our attentions toward many directions at once, but such a demand is the charm of any fair. First, we shall see more of the novel ideas spoken about. Then, we shall attend to the Midway's display of native peoples; not images presented by paintings or sculpture, but living, functioning, unpredictable, real people. Then, after our respite, we will return to the exhibits proper, and examine the exhibits of three budding sciences.

The World's Congress of Ideas

A byproduct of the fair celebrated 'man's achievements' by bringing to Chicago the great teachers and thinkers of the times. [2] The intellectual demonstrations were arranged through what was called the World's Congress of Ideas, for which many academic organizations agreed to hold their annual meetings in Chicago at the time of the fair. To supplement these, the fair organizers arranged talks, sermons, and lectures. Using the new University of Chicago and the Institute of Art, the World's Congress Auxiliary provided, from May 15 onward, an intellectual feast that followed the tradition of Paris a few years earlier, at which symposia were held on the subjects of hypnotism and magnetism and the new experimental psychology. Chicago provided 5,978 speeches, of which 3,817 were spoken and the remainder added in print later. A single place and time had not seen such a gathering of notables since, perhaps, the agora at Athens in the fifth century BC, or the head-choppings at Paris in 1798.

The motto of the World's Congress Auxiliary was 'Not Things, but Men.' Among the academic groups that sponsored meetings were the Congress of Women, the Temperance Society (a gutsy locale for them), Medicine and Surgery, Commerce and Finance, the Literature Society, the Modern Language Association, the American Philosophical Society, and the American Medical Association. The meetings of the The World's Parliament of Religion, which continues to meet into our times, dwarfed any previous such gathering and included Annie Beasant, of whom we will shall learn more later.

At the fair, Woodrow Wilson, still president of Princeton, if also a politician, lectured on why lawyers, physicians, and clergy should receive a liberal arts education before undertaking their professional training. Frederick Douglass noted that the opening ceremony featured no representative of the "Negro classes" except in the 'cultural displays'. Mrs. Potter Palmer, proponent of the Women's Building and one of only two women who spoke at the opening ceremonies, lectured on women's role in a just and productive society. Dwight Moody, supplier and interpreter of bibles, worried whether it was honorable to attend an Exposition that was open on Sundays. (At first, the Fair was not, but later was opened on the Christian Sabbath, in deference, it was explained, to the many religions represented, if not to the gate receipts.) The Labor Movement secured permission for an open-air meeting at the lake -front at which Samuel Gompers and Clarence Darrow spoke of the evils of concentrated wealth. Police and workers, or police and anarchists, depending on one's view, had killed one another seven years earlier not far away in the Haymarket. But President Grover Cleveland, who opened the fair by pushing a button that started the electricity that unleashed the machinery, fountains, and lights, refused to pardon the workers/anarchists convicted of murder during the still unsolved Haymarket riot.

Other speakers, troubled by the distribution of wealth, argued for a graduated income tax, public works to create employment, a six-day work week for adults, abolition of children at heavy work, the establishment of state-run schools with compulsory education, and the futility of violence compared to 'peaceful cooperation'. If the fair's displays looked backward, the intellectual and political involvement of the many such groups in attendance looked forward.

Psychology was represented by two separate symposia: One, on experimental psychology, was chaired by G(ranville) Stanley Hall [1844-1924]. The other, on 'rational psychology', was chaired by former Princeton President James McCosh, who so had so valiantly tried to unify evolutionary and religious principles. This would be McCosh's last public appearance before his death, at age 82. He read a paper entitled "Reality: What Place it Should Hold in Philosophy". The Fair was to be of importance to that towering intellect, the German scientist Hermann von Hemholtz [1821-1894] whose researches on the physics of the sensory systems, especially the eye and ear, gave experimental psychology a starting place. The cold he caught at the Fair turned into far worse problems, these leading to his death from pneumonia on his return to Germany.

Figure 3.2: The fairgoers. Top: the three persons to the right were probably dressed for their appearances in the displaying of the advance of human culture by the presentation of racial and cultural groupings.. Bottom: a mix of fairgoers at the Midway, wherein 'fun' attractions gave respite from viewing the serious displays and attending the Congress of Ideas.

The Midway

The central strip of the fair featured the giant Ferris Wheel, the first constructed. As its coaches arose 250 feet above Chicago, one could experience a 'bird's eye' view of the fair grounds, the lake, and the city, in a way never before possible save by balloon ascension or by ascending building that used the recently innovated iron architecture such as the Eiffel Tower built for the Paris Fair only four years earlier. There were displays of home industries (the making of Irish lace), food and drink (French cider), adventures (one could kiss an ersatz Blarney stone), glass blowing, the native American Indian Sitting Bull's cabin, an ostrich farm, and Jim Corbett demonstrating his defeat of John L. Sullivan in fisticuffs. [3]

It was in this area, the Midway that the folk acquired for the living- anthropological exhibits lived and performed. There were also the exhibits of deformed folk, food, and exotics that grace all such fairgrounds. The Javanese males performed traditional songs and dances, the Laps showed their reindeer (Caribou to North Americans), and even in the summer heat of Chicago, Eskimos showed off their fur wraps. We are told [4] that it was the dancing girls from North Africa who inspired most attention, perhaps because the demonstrations of included Little Egypt performing a 'danse du ventre,' also known as the 'belly dance'. At the time, these movements were somewhat coyly described as "She does with her belly what others do with their feet". Thirty-five years later, her gyrations were specifically included in a scene based on the fair in the musical *Showboat*. Tribal marriages were performed, children reared and, obviously, romances were begun, consummated, and concluded. The police record shows a total of 954 arrests, and the finding of three fetuses. [5]

The folk displayed at the Chicago Midway-Plaisance were reported to have accepted the assignment — whether out of an eagerness to leave their homeland to see Chicago, to display their culture, or merely to experience adventure, we do not know. At Chicago, the displayed peoples were not captives, in the sense that they were not captured, so far as we know, nor were they jailed, although pressure was surely used to keep them in the fair's campgrounds. This, it was argued, to maintain the integrity of the exhibit. St Louis's 'Louisiana Purchase Fair' of 1904 would do the same, for the last time on such a scale.

'Sets of living people' were exhibited to reveal patterns of race and culture. Displaying living people led to difficulties, both in acquiring such and in caring for their needs during their long stay within the fairgrounds, but such was managed. We know from the report of the governing committee that the fair organizers sent a scout to Africa to 'hire', but the very extensive records of the fair say nothing of techniques of employment or the contracts made for the Africans. We know even less about how the Asians were recruited. When this gathering was finished, the native groups offered for the edification of fairgoers were families and groups of American Indians from North and South America, Eskimos, Egyptians, the Dahomey (perhaps from Africa, or possibly, as the musical *Showboat* suggests, from the New York City's Avenue D), Hawaiians, Javanese, Samoans, Algerians, Bedouins, Persians, Japanese, Chinese, and Muscovy. Dance and music were featured, but some groups, the Dahomey being conspicuous examples, took seriously the charge to display their customs not only as represented by music, but by presentations of family customs, including marriage, and preparations for war. There are hints that they took special pleasure in the war chants, as these seemed to have the desired effect on the largely Caucasian audience.

There is a persistent belief that the Midway's displays of peoples were organized according to some presumed path of evolution - a sort of ladder of civilization ranging from headhunters to Americans of northern European descent. Such was not the case. The displays of peoples were laid out by latitude, from the northern -most peoples (the Esquimaux) to equatorial peoples, these being from the south, at least by northern standards. The difference in placement may of course have been disingenuous, as many surely held the still prevalent belief that, except in Great Britain, northerners are intellectually and physically superior to southerners. Regardless, it is quite likely that the average-fair goer saw the displays as arranged in the order of human progress. [6]

It is commonly supposed, but untrue, that Darwin astonished the world with the publication of his theory of evolution, much as Freud, a few years later, was to awake the world with the notion of the 'unconscious'. Certainly, there were those in Darwin's time, and those living today, who deny that living material evolved into its present form through natural selection. However, Darwin's ideas were not entirely novel. One author of a popular article on garden flowers had written the connection and later claimed credit; other, more academic, less practical folk, had nibbled at the central idea without catching its center. Darwin's major work on evolution sold-out within hours, this either a publisher's ploy that would make modern publicists swoon, or evidence that Darwin was writing to a populace already awaiting his analysis. By 1893, the 'theory' of evolution was known as the 'Darwin-Wallace theory' thus honoring Alfred Russsel Wallace's letter to Darwin outlining the fundamental idea. (Wallace, generously, had put an end to this by writing a book in 1889 called simply *Darwinism*.)

The voluminous accounts of the exhibits, programs, and records made by the organizers of the Fair further support the idea that Darwin's views were accepted by educated folk if only in their most general terms. The discovery of dinosaur and large mammal remains in the American West, a cause of much excitement in the 1880s, gave Darwinism a presence that was very real and newsworthy for Americans. If the Chicago fair-goer was a little confused about humankind's descending to, or ascending from, the apes, presumably with aborigines showing the way, what was understood to be Darwinism was generally accepted. Every there were signs of progress, of goals met, of goals to be achieved, of half and quarter-way steps between and among people and animals, of humankind's ability to fashion and re-tool the world for (his) convenience. Variation and gradation were everywhere in evidence and, as arranged, were clear grand examples of God's design.

Here we find a small but promising mental fossil. To the minds of the fair-goers, God had a goal in mind, and that goal could be determined through observation. If the Divine Hand was not always guiding visibly, it was doing so invisibly, as Adam Smith had noted was the case in capitalism. The notion of truly random effects, that the design of creatures and their activities were purposeless, seemed wrong-headed. Today's view of the course of evolution, at least that held by some, is that it is purposeless: What we see as design and purpose is thought to merely be our mind imposing order on otherwise chance occurrences. But to the mind of a century ago evolution was understood as evidence of God's influence, and was fully compatible with belief in a purposeful universe.

We have no direct justification by those involved in the procurement as to how the mind of this time and place was able to conceive of the idea of human evolutional displays, nor how the plan was carried out to display the cultural and intellectual progress of human beings. We wonder of this just as generations future to our own will be interested, perhaps, in grasping why folk of our time are conflicted about the propriety of maintaining zoos, not to mention animal-shoots (events in which large numbers of caged animals are released before a crowd equipped with guns).

The displaying of human beings presumably living naturally in their cultural settings was not unique to the 1893 fair, although such displays were most likely initiated there. We shall use our ability to skip to the future to visit three men who themselves would be kept in settings for education and entertainment:

Let us visit Ota Benga, Minik and Ishi, for their stories may help us understand the beliefs of those who created and viewed such displays. Our contemporary views ascribe different feelings to the conditions of these men. I believe that there is a general feeling today, held by those familiar with one or more of the stories, that Ishi was fortunate, his treatment humane. In contrast, we read of Ota Benga with remorse, viewing his treatment as

63

characteristic of the thinking of another time, another place, another kind of humanity. Minik's story arouses sympathy for him, distrust and dislike of those responsible for his transfer and life in America. But these feelings are our modern view. For different purposes, all three were on exhibit or made their homes in the United States at the same time, the first and second decades of the nineteenth century; one at the Bronx Zoo, one at the American Museum of Natural History in New York, and one at the Anthropological Museum, then in San Francisco.

Living Human Displays Beyond the Fair

The people whose lives we meet were chosen by their captors and keepers because they represented distinct variation, perhaps a representative of a race and culture unfamiliar to the examiners; perhaps someone thought to be the last of their race, therefore providing the last chance for information about the group to be gathered. Perhaps they had nowhere else to go except into a society that wanted to examine them. They were chosen due to the fascination with variation, and the belief that variation could be seen, studied, and cataloged. Photography, as we shall read, made the recording of variation possible and the development of tests and measures of human variation appeared a progressive way to predict skills and talents. What the displayed people have in common is their physical, cultural, perhaps mental distinction from those who became responsible for them, a form of colonization. Least we moderns think such exhibitions to be the oddity of another age, consider Figure 3.3, a contemporary photograph, showing a young visitor inspecting bushmen children at an 'educational' display.

Figure 3.3: photo ©New York Times. At a contemporary park wherein attendees can examine native Bushmen.

The first of our displayed human beings is Ota Benga (Figure 3.4). He was 'attached', 'employed', or 'exhibited', depending upon one's perspective (7), at the turn of the century at the Bronx Zoo after his 'liberation' (or capture) from the Congo. He resided with the New York Zoological Society at the Bronx Zoo and later developed a second career, this doing dances and the like at the 1904 St Louis Exposition and elsewhere. He was promoted as a display, for example, as a piece d'resistance following the annual dinner of the New York Zoological Society attended by Caucasian males in white-tie evening dress. He was a celebrity, and seems to have been used by the director of the zoo much in the way that recently acquired exotic animals are advertised by today's zoos with the intention, they say, of educating the public; increasing the gate must be an added bonus.

Figure 3.4: Ota Benga, housed at the Bronx Zoo, with a companion.

There is far more to note about the Eskimo Minik, who was removed by Admiral Peary from his home in Lapland along with five other members of his village on arrangement with the American Museum of Natural History and a New York newspaper. He was the only one of the group to survive their stay in New York. Minik was manipulated by various people for various ends, but ultimately found friends. Following his exhibition at the Museum, and the death of his companions and father, Minik was enculturated into American civilization not by policy, but by the kindness of others [8]. He was befriended by an administrator at the Museum, Mr. and Mrs. Wallace (unrelated to Alfred Russel) and their son, whose surname Minik took. He also took Peary as a middle name. Education was sought for him, and he then lived in New York in a hotel on West 43 St. off Times' Square.

Like many tourists, Mink eventually tired of New York and city life. Mink returned to Lapland, married, fathered, and, then tired of Lap life, returned to New York. He then gained nation-wide publicity when he claimed to have discovered the bones of his father on display in the museum. His surprise originated from the fact that after his father's death, Mink, then a boy, was taken to the funeral shown the coffin before it was lowered. Admiral Peary, then one of the most celebrated people in America, when asked if he had a responsibility to the young man, showed little interest in the hegira of his captive: he referred to Eskimos as 'animals with speech' an odd assessment as, on his journeys to the Arctic, Peary fathered at least two children with an Eskimo women. Both children lived into adulthood and were acknowledged, if only slightly, by the Peary family.

Minik made public his story of the treason regarding his father's bones, and his complaint that he had been deceived was taken in ill-humor by the director of the Museum. He became non grata, and, rather like an animal involuntarily contributing to human experimentation whose usefulness is ended, the policy of the Museum was to say as little about his whereabouts as possible. Minik's encounters with civilization, and especially with humane and warm members of this family are best left to each reader to uncover and assess, and there is no better way to do it than to read Kenn Harper's book *Give Me My Father's Body*.

The life of Ishi, the last living native American Indian of the Yana tribe, and surely a truly native Californian, who found his way into northern California Caucasian civilization in 1911, is apparently of a different nature. He was provided the guardianship of anthropologists from the University of California (Berkeley) who provided shelter, food, and companionship, and who investigated his language and recorded his arts of hunting and tool-making. Ishi lived the last four years of his life in the San Francisco Museum of Anthropology. His knowledge of native-American culture and language meant that he was able to provide information that otherwise would be lost upon his death, such as how his tribe made arrows and flint, what they ate, and their other customs. However sensitive the staff was to his needs and feelings, then as now the press knew how to milk a story. His reactions upon being taken to a stage production were recorded by the San Francisco Sunday Call on October 8, 1911 in an article that tells us something about Ishi, to be sure, but also some things about how native-Americans were then understood.

"With broad shoulders squared, head bravely thrown back and eyes somber with fear and wonder he pussyfooted down the aisle of rich plush. Ishi, the primordial man, the only really wild Indian in existence. the last of the cavemen took his seat. . . At his side were learned pundits. . . almost touching elbows with the saddle-colored primordial man were gentlewoman of the conquering people, soft-voiced and beautiful, their lithe shoulders agleam with flashing jewels. . . . Cold terror sat upon him at first, but terror bravely mastered and hidden under a mask of stoicism such as only a son of the wilderness may wear. . . . Never before had he seen white people, excepting in small groups. He could not believe there were so many people in the world, and knowing nothing of paleface custom, save what he had seen once, 40 years ago, when the gold seekers had slaughtered practically all of his tribe before his eyes, it is small wonder that he misjudged the spirit of vaudeville. To him the stage was the mystery room of the gods, the singers were priests, the dancers were medicine men and women, and the orchestra was designed to drive the devils out of sick people. . . Later he asked the interpreter whether the applause helped to drive the demons away, as he had observed that everybody ran off the stage when the people spitted their hands together." [9]

Douglas Keith Candland

Figure 3.5: Minik, dressed in his Labrador wear, poses for a publicity
photograph while he was living in New York. Reqquested from Kenn
Harper and Steerforth Press, 1986, 2000.

Figure 3.6: Minik, enculturated and living with the Wallace family near New York City. Requested from Kenn Harper and Steerforth Press, 1986, 2000.

Figure 3.7: Alfred Louis Kroeber, documentor and friend of Ishi, with Ishi, here dressed in the 'civilized' style. Requested from Kroeber and the University of California Press, 1961.

His relationship with the California anthropologists Alfred Louis Kroeber and T. T. Waterman, who first interviewed him in Oroville, California [10,11], has many resemblances to the descriptions of supposedly feral children who found their way into civilization in the eighteenth century, thereupon to serve both as objects of scientific investigations and of the kindnesses supplied by caring families and helpers. True, a research specimen Ishi was, and his demonstrations of tool-making, canoe-making, hunting, and the like were surely of cross-cultural significance, while his place of residence, the Museum in San Francisco makes it hard to argue that he was integrated into society. Nonetheless, he was shown respect and cared for, regardless of motivation, this being yet another common justification for captivity.

From the view of one of the anthropologists, T. T. Waterman:

"He convinced me that there is such a thing as a gentlemanliness which lies outside of all training, and is an expression purely of an inward spirit. It has nothing to do with artificially acquired tricks of behavior. Ishi was slow to acquire the tricks of social contact. He never learned to shake hands but he had an innate regard for the other fellows existence, and an inborn considerateness, that surpassed in fineness most of the civilized breeding with which I am familiar." [12]

Figure 3.8: Ishi demonstrating the use of the bow and arrow. Requested from Kroeber and the University of California Press, 1961.

Current writings treat Ishi's Caucasian companions as compassionate. His presence was unarranged by people of the dominant culture (unlike those of Ota Benga and Minik), but once he appeared, persons of science and medicine requested his presence, ordered his activities, and gave him home and work in the San Francisco Museum. A federal government agency, the Bureau of Indian Affairs, watched the process and offered him sanctuary elsewhere, chiefly on Indian reservations. This fate, Ishi declined, preferring to work and live in the museum. His work consisted of displaying his tool-making abilities and something of Indian life to visitors at the museum. If not on display, he was, surely, engaged in displaying himself and performing Native American tasks for the edification of visitors.

Ishi was invited to Caucasian affairs, dinners, plays, musicals, and such. His reactions, say to women, were observed and recorded. He accompanied his Caucasian 'chiefs', as he called them, on camping trips to Ishi's native territory where they hoped to observe other techniques characteristic of Californian Indian tribals. In addition to the scientifically-oriented monographs produced by the chiefs, Theodora Kroeber, wife of the anthropologist Alfred Louis Kroeber [1876-1960], who took the earliest responsibility for Ishi, wrote a book which, in several versions, describes the situation that resulted from Caucasian settlers taking over California Indian land, Ishi's life in captivity, and his treatment. The text is intelligent, far-ranging, informative, yet necessarily colored by the relationship of the author to a principle in the story. The text does not discuss this relationship, but the dedication makes it clear. Her sympathy with Ishi, her tact in describing his beliefs and behavior, yields a book that is both complete and warm. Generations of California students have read this book, and it is fair to say that Ishi has become an icon of the 'savage meets civilization' in literature.

In the end, Ishi is something of a tragic hero; a complex Shakespearean figure, to us. He was acculturated, his culture studied; he was integrated into Caucasian society in a kind way, yet he was a specimen; Ishi and his Caucasian companions summarize, perhaps, the puzzles that surround captivity. Each question we ask reveals a deeper one; no solution presents itself, while Ishi's pre- and post-Caucasian life can be interpreted to suit any number of ethical views and justifications. Perhaps the most striking statement about Ishi comes in an obituary of his life written in the (California) Chico Record of March 28, 1916:

> "He could not stand the rigors of civilization, and tuberculosis, that arch-enemy of those who live in the simplicity of nature and then abandon that life, claimed him. He furnished amusement and study to the savants at the University of California for a number of years, and doubtless much of ancient Indian lore was learned from him, but we do not believe he was the marvel that the professors would have the public believe. He was just a starved-out Indian from the wilds of Deer creek who, by hiding in its fastness [vastness?] was able to long escape the white man's pursuit. And the white man with his food and clothing and shelter finally killed the Indian just as effectively as he would have killed him with a rifle. "[13]

As was said at the beginning, sometimes a discovery in one location leads to digs in other unexpected locations. We return, however, to our main task, and continue to uncover the fossils available to us at the Columbian

Exposition. In Chapters 6 and 7 we shall read how persons in authority (physicians, professors) measured the naked bodies of college students in the hope of predicting behavior from body-type, while lawyers and professors worked in somewhat like fashion to predict which adolescent children would become criminals.

Department M

The divisions of exhibits at the Fair reflect the classifications of human achievement: agriculture, mines, fisheries, food, clothing, decorative arts. Divisions of a social and intellectual nature also complemented the Fair's attempt to demonstrate the degree and diversity of human nature and achievement. A department of "Science, Religion, Education and Human Achievement" was proposed but was later annexed by the Department of "Liberal Arts and Education." This department was especially active because of the academic and professional organizations holding their annual meetings.

Department M, the last to be organized, was "Ethnology, Archaeology, Labor, and Women's Work." This housed the psychology, neurology, and anthropometric alcoves, which exhibited the scientific progress of the fields including and the measurement of individual differences, capacity, gender, and race. Their importance is that they were the first public presentations of the achievements of three new academic departments. Exhibits for each occupied small alcoves along one wall. The floor plans [14] show these as each occupying three 12 by 12 foot areas of the building which was itself 415 by 225 feet. Small areas, but the beginning of recognition of three fields of study that would shape human beings' understanding of themselves in our times. The contents of these exhibits are valuable mental fossils both for what they expressed of the times and how they would develop in the future.

One of these alcoves offered an exhibit of the techniques and achievements of the emerging science of anthropometric measurement. Included were displays of German and French instruments specially designed for the budding enterprise. [14] The display of anthropometry emphasized the use of photography to record physical variation and suggested that physical aspects of people are correlated with their mental and behavioral traits. A practical use of such would be the identification through body or facial characteristics of those practicing, or likely to practice, criminal behavior (Chapter 7). A like idea, sometimes discredited, sometimes not, was the idea that intellect, talent, and character could be predicted from certain measurements of the head (Chapter 5) or the shape of the body (Chapter 6).

The motives of the anthropologist of the day are often evident in their writings: some wish to establish the evolution of the races, some wish to correlate body and cranial size to individual ability, others begin with the view that races are differences in ability, and still others seemed happy to have file-cards of measurements. It should be noted that the term 'racial', used at the exhibit, as in 'racial psychology', and 'racial anthropology', specified an academic discipline and was not necessarily an academic pejorative; that there were intellectual and physical differences among and between races was assumed.

Figure 3.9: Sample exhibit at the Chicago Fair of 1893 on facial types and criminality. From the exhibition of, perhaps, psychology or anthropology. Permission requested from Brown and the University of Arizona Press.

A major task of anthropometry at the time was to use bodily measurements as indicators or correlates of temperament. To this end, the Harvard University medical and physical education authorities had been photographing enrolling students, unadorned. An example is shown in Figure 3.10. Each student was photographed from three angles so anthropomorphic measurements could be made of aspects of the human form. From these measurements were sculpted the figures of the 'representative' man and woman, dubbed:

'Adam' and 'Eve', arranged tactfully for exhibit, as shown in Figure 3.11. So tactfully, in fact, that on the night before the opening of the fair, the authorities insisted that Adam's body be turned to the wall. Perhaps the measurements and reconstructions were far more thorough than contemporary sources describe. If measures were taken from the entering classes at Harvard and Radcliffe only, they are based on American Caucasians of north European descent, although it is not evident that the legendary Adam and Eve were of this genetic and social group.

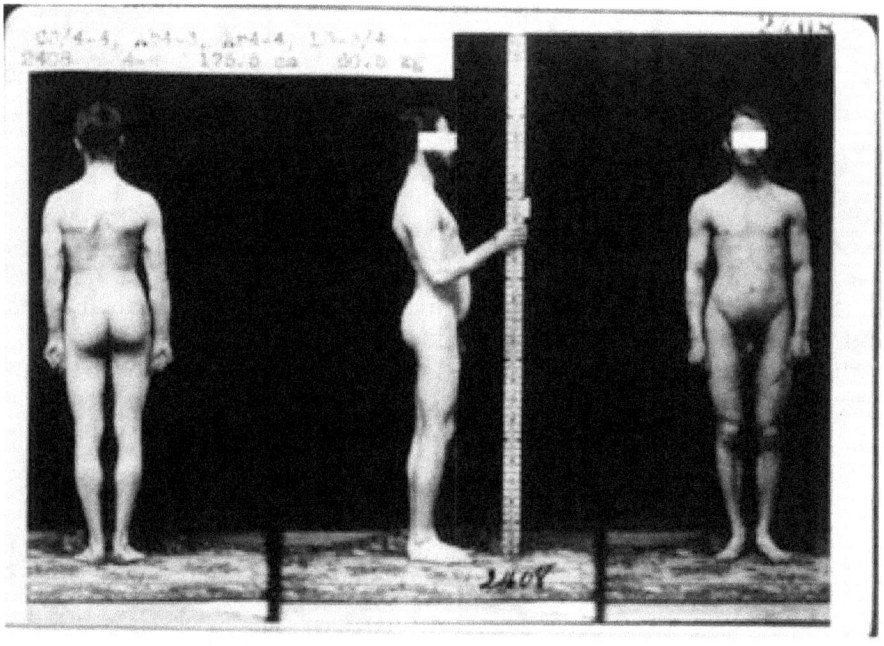

Figure 3.10: An example from the Harvard study of the physique of male students, an idea reflected later as 'constitutional psychology' in which prediction of men (and later, women) was made of temperament and likely criminal behavior. Permission requested from Ambler, Banta, and Harvard University Press.

Figure 3.11. The perfect man and perfect women, as calculated as measurements of students at a university, displayed at the Chicago Fair, 1893. from J. K. Brown, 1996, p. 45. Permission requested from Brown and Harvard University Press.

The photographing of persons unadorned as part of their college-orientation has been treated by contemporary re-discoverers of it in several ways. [15] Some have asked, how is it that young people were willing to accept these orders, to allow themselves to be photographed naked, supposedly for the betterment of their bodies and likely benefits to society itself through the data they provided? One answer lies buried within the question itself. In that era, it was believed that social science would improve the human lot; that research on behavior and society could reveal something about how society might end or ameliorate human suffering. Cooperating with the investigative needs of social science was, in a sense, patriotic. The idea of 'human betterment' through 'human progress' guided acceptance of what was a perhaps slightly stressful, but hardly life-threatening, duty.

A second exhibit, this displayed in a small area of the same hall, featured equipment representative of the new laboratory psychology. No doubt, the presenters were proud of these demonstrations of how to measure variation with accuracy. Such laboratory-based psychology as there was in the United States had been emanating and was transferred from physiology (in Germany), empiricism (in the United Kingdom), and religion and 'mental health' (in the United States).

Of the exhibit as a whole, we learn:

"In these [displays] can be seen a large collection of instruments and apparatus, received from the more important anthropological laboratories of the universities in this country and from several in Europe with a very extensive series of apparatus from the principle makers in Europe made especially for this exhibit. The laboratories are divided into three sections — Physical Anthropology, Neurology, and Psychology." [16]

In these laboratories the practical working of the apparatus is shown and any one who wishes can have, by the payment of a small fee, various tests applied and can be measured and recorded upon cards which are given to the person, while the record is made upon the charts and tables hanging on the walls of the laboratory to illustrate the various subjects. [17] There too, are a series of skulls and skeletons and various models showing the physical characteristics of the various races and varieties of man. An interesting series of charts in the Physical Anthropological section is that illustrating the development of over 50,000 school children in various cities in North America; while another series of diagrams and maps shows the physical characteristics of the Indians of North America, as derived from measurement and observations upon nearly 20,000 Indians, recorded by about twenty-five special assistants of the department, who were engaged for nearly two years in this work. [18]

William James [1842-1910], whose textbook on psychology, published two-years earlier, successfully redefined and translated European epistemology into American psychology, did not attend because of conflict with his annual summer trips to Europe to catch up on academic progress and culture. Perhaps he was tired of the fair scene, as he had attended and conducted seminars at the Paris fair only a few years earlier. However, had he attended, he would have noticed a difference between the fairs. The academic work featured in Paris was on animal magnetism, hallucinations, and hypnotism, while that displayed in Chicago showed the cusp of the shift from the medieval and spiritist to modern experimental psychology. He sent, instead, parts of his fledgling psychological laboratory at Harvard, displayed by assistant professor Hugo Munsterberg (1863-1916). Munsterberg's reputation would be enhanced later by his studies of mediums and spiritism, a topic that re-appears in Chapter 4. Joseph Jastrow (1863-1944) of the University of Wisconsin is reported to have headed, with the hesitant approval of the young

American Psychological Association, Galton-inspired measurements of those willing to pay a 10 cent charge. Measures were made of individual's physical attributes as well as mental ones in the hope of creating a large sample of the averages and statistical deviations of the US population.

Jastrow and Munsterberg were both to become influential persons in the use of psychological principles for practical affairs. The relationship between academic psychology and commercial psychology was soon to be established, this being an industry that separates the motives of social science into our day. The professors, more interested in pragmatic issues than James, demonstrated the individual differences to be found in the reaction time of the nervous system, along with other promising techniques involving psychological measurements. The idea was promoed that such tests and measures could be used to distinguish employees' abilities

The similarity between these measurements and those of Mr. Phillips' skull does not escape our eye. The testing-movement in North America can be said to have begun at the Fair. With impetus from World War I, when tests were used to distinguish the likely abilities of soldiers, along with the development of tests of intelligence, aptitude for college, and the like, testing would soon become the most influential and widespread tool of psychology and, perhaps, of all social science. [19]

At the Chicago Fair, the Harvard Psychological Laboratory was duplicated. Equipment available at the time was mostly of German design and construction. These included the Hipp Chronoscope (for measuring small periods of time), the Hempschluge kymograph (for tracing events over time), the Gedachtnisapparat (for studying differences in memory), Marbe's Color Mixer (for studying how hues combine in the retina and brain), the Galton whistle (now known as a dog whistle, because of its high pitch, used to measure differences in auditory acuity), and a Mirror Rotator, this to induce hypnotic states. [20] J. Mark Baldwin [1861-1934) one of James McCosh's 'new men' at Princeton, was judge of psychology exhibits at the fair, and describes to us the purpose of these exhibits:

"The Psychological Laboratory — The object of this laboratory is to illustrate the methods of testing the range, accuracy, and nature of the more elementary mental powers, and to collect material for the further study of the factors that influence the development of these powers, their normal and abnormal distribution, and their correlation with one another. The laboratory is thus designed, not as are those connected with universities, for special research, or for demonstrations and instruction in psychology, but as a laboratory for the collection of tests. As in physical anthropometry the chief proportions of the human body are systematically measured, so in mental anthropometry the fundamental modes of action upon which mental life is conditioned are subjected to a careful examination With this determined, each individual can find his place upon the chart or curve for each form of test and from a series of such comparisons obtain a significant estimate of his proficiencies and deficiencies. The problems to be considered are such general ones as the growth and development with age of various powers; what types of faculty develop earlier and what later; how far their growth is conditioned upon age and how far upon education; again, the difference between the sexes at various ages, differences of race, environment, social status, are likewise to be determined. The relation of physical development to mental, the correlation of one form of mental faculty with others, the effect of a special training, -these, together with their many practical applications, form the more conspicuous problems to the elucidation of which such tests as are here taken will contribute. In addition to the interest in his or her own record, the individual has thus the satisfaction of contributing to a general statistical result." [21]

Notice that most of this equipment was designed to test and measure differences among people. Thus the psychology of this display advertised the experimental approach to human and animal variation. This fossil motif was the legacy of the nineteenth century's discovery of the importance of focusing on variation. For example, consider reaction time, the time required for a human being to make a motor response to a perceptual stimulus, which was originally thought to measure differences in the speed of nervous-system transmission. But as variation was found and its significance recognized, the practical aspects of reaction time in predicting success at work, intelligence, and the like, became utilized. It seemed obvious and commonsensical that the quicker one transmits ideas and instructions, the 'brighter' and more capable one is apt to be.

These were not the only displays of psychology and anthropometry present at the fair. The German Educational Department and the Deutshe Gesellschaft für Mechanik und Optik each had displays of the psychoholgical research instruments, though we are told a collective display of German makers at the north end of the Anthropology building is far more complete. French companies also displayed their surgical, physical, and psychological instruments, though they did so individually. As to the differences between the displays, we are told that "The German makers have done their work more largely in connection with great university laboratories, and so have subserved better the needs of particular students in solving particular problems in physics and psychology: the French, on the other hand, have found the demand more marked from the side of clinical medicine and experimental physiology." [22] Further, there were two exhibits by universities. The University of Pennsylvania exhibit, located in the Liberal Arts building, was intended to present an entire working laboratory and active data collection, but due to a lack of staff merely displayed tests of reaction-time and visual aesthetics. The display of the University of Illinois, located in the Illinois State building, displayed the collections of the Department of Ethnology.

The idea of the potential of measurement is evident; less apparent is that of photography, the mechanism by which images are seemingly forever set. We should note here in passing that, like electricity, photography has the characteristic, to the observer, of being an unseen power.

Electricity and Photography

The two remaining aspects of the Fair that command our attention are the use of electricity and the process of photography. Contemporary accounts of the fair repeatedly turned and returned to expressions of awe regarding the electric lights. If "awe" is the correct way to describe how electricity was regarded, "pride" is the proper word to describe the then contemporary descriptions of the many uses of photography among the exhibits.

Electricity as an example of how technology becomes translated into metaphor and simile about life itself. If electricity was what grabbed and held the awe of viewers, so, it seems to me, may electricity be the 'root metaphor' of the mind that guided the intellects of the times. It, may I say, 'illuminated' the possibilities of human progress. It promised night's becoming useful and far less dangerous; some had even suggested that it might permit messages to be transmitted rapidly through the ether, although this seemed unlikely, rather like some of the quaint 'spiritist' ideas described in the next chapter. But, there was something spiritistic about electricity: for electricity offered evidence of activity in unseen worlds, of imperceptible processes, of serious and powerful forces whose origins were obscure and unobserved. Who would guess that something unseen could be so predictable, and so useful? And, if electricity was unseen, what other unseen spirits awaited discovery? Writing in the era of the Chicago Fair authors stumble time and again to find words, this in an age of extravagant writing, to describe the effects of the electric lights.

Our explanations of ourselves, certainly of how our brains work, appear to follow the metaphor provided by the technology of the day. The sixteenth century in Europe imaged the brain to work rather like the water-driven figures at Versailles, an image that Descartes used. Later, Mesmer and others thought the mind worked through the phenomenon of magnetism. Most in the 20th century offered an explanation of mental function which featured power, resistance, and outcome, reflecting the new understanding of electricity. The best known of these is Freud's model of the mind that imagines power (id) to be constrained by resistance (superego) yielding ones grasp of the world (ego). The connection between id, superego, and ego is, metaphorically, identical to Ohm's law regarding the properties of electricity. The second, that of Konrad Lorenz in ethology (a science of animal behavior), suggested that 'fixed patterns' of behavior could be released by 'innate releasing mechanisms' in a manner that appears parallel to the theories of Ohm and Freud. The third was the belief in unseen worlds, this known as spiritism, from which human behavior, past and future, was guided. Whatever technological marvel is at hand when we construct a theory of the brain and behavior, but let us not digress that far.

By the time of the 1893 Fair, photography, of course, had been around for sixty years or more, and it was no longer a magical oddity. Yet only now were its practical uses being extended to academic science, especially to art, anthropology, neurology, and psychology. The exhibits are surely influenced by recently invented techniques of photography. The still picture opens up not merely a way of communicating visual images, in the sense that we are able to "see" other peoples, buildings, landscapes: more, we have from photography, documentation that can be stored, evaluated, and even re-evaluated when other needs are recognized. It provides a more or less permanent record for that which, before, could only be described in words or artistic impressions. Some aspects of Anthropology would come to see their basic data as being entombed in the photograph, rather like a fossil. And the photograph would become an essential tool for those wishing to demonstrate correlations between physical aspects of the body and

89

talents; delinquency and criminal behavior among these traits.

It was understood that the scientific use of photography was not merely a visual record taken from one angle or another. The value of the photograph was in its ability to record from different angles and to display thereby a sense of depth and surely a sense of proportion to the two-dimensional image. So it was that the photographs of criminals by Alphonse Bertillon in the 1880s and 1890s, these shown at the Fair at the French pavilion, display the criminal 'head' from different angles, not from the front view alone. This encouraged measurement, naming, and categorization, which Aristotle told us are the beginning procedures by which science organizes its knowledge.

Both psychology and anthropology found immediate use for the photograph. Darwin [23], in his 1872 book on emotion, uses photography to document data critical to his theory of human and animal emotion. He had photographed facial images of the insane, his children displaying emotional states, and of the facial expressions of actors miming emotional expression. His experimental interest was whether other folk could identify the emotion being expressed or emulated. Usually they could, thus suggesting to Darwin the universality of facial expression. Contemporary research that has both refined the methods and enlarged the peoples from whom expressions are recorded substantiates Darwin's tentative conclusion.

Galton's Shadow: Sir Francis Galton and the Measurement of Variation

We must pause here for a moment to highlight the work of one man, a portion of which was displayed at the fair, and who's other works will come up later. Recall Francis Galton, Charles Darwin's half-cousin, who was among the most distinguished men of his time. Galton was amoung that group of British who, given inherited wealth and position, could devote all of their time to their interests in intellectual and academic pursuits. Among his many lasting discoveries were the statistical tools that form the basis of social science, chief among these being the measurement of statistical correlation. [20] He created many of these techniques in order to determining the physical and mental capacities of his fellow Englishmen as he believed that one could not hope to establish how evolution worked without having baseline statistical descriptions.

Galton believed that humanity had evolved through natural selection, that the differences among races and cultures were self-evident, and these states could be arranged along a ladder or continuum showing the evolution of mankind. It was axiomatic that the English, as distinct from the British, sat on the pinnacle. But even within the category 'English', evolution's effects were to be found, as the differences among classes of people were also evident. Working-class people over-reproduced, while the upper-classes did not always do their duty to reproduce more heavily. Education should not be wasted on those who would not profit from it, Galton believed, and those of superior talents should be given special attention. These interests in heredity and class led him to collect measurements on how people differed mentally and physically. His purposes, he tells us, were to understand the qualities and composition of the British people, by which he seems to have meant those living in England. He explains, "We want to know all about their respective health and strength and constitutional vigour; to learn the amount of a day's work of men in different occupations; their intellectual capacity . . . the dying out of certain classes of families, and the rise of others. . . in order to give a correct idea of the present worth of our race, and means of comparison some years hence of our general progress or retrogression" [24]

To launch the collection of data, Galton wrote to headmasters, requesting a school's science-masters to measure the height and weight of pupils. Two schools replied, and with the addition of data a few years later, Galton was able to publish information on the averages from city (private) schools as contrasted with rural (public) schools. Galton found differences, for example between rural and city fourteen-year old males: the rural group was taller by an inch and a quarter, and weighed more, by seven pounds.

If we pause for a moment to ask what the differences mean, we get nowhere quickly. We cannot say that such differences are merely a matter of city versus country living, for certainly location co-relates with wealth, education of parents, medical care during development, nutrition, etc. Nor was Galton able to say whether the differences found were sufficiently large to be statistically significant, for inferential statistics were yet to be invented. Nonetheless, for his time, Galton's research was as advanced and scientific as any (and perhaps more so than some research which is done today).

Galton's interests led him to participate in an International Health Exhibition held in South Kensington, London, in 1884. [25] Here visitors could move along a passageway along which were arranged instruments for measuring aspects of their minds and bodies. For a fee of three-pence, visitors were measured by Galton and two assistants. The visitors received a card showing their measurements and comparing them to the mean, while Galton kept a copy for his researches. Among the measurements taken were "weight, sitting and standing height, force (hitting an object), reaction time, keenness (thresholds) of sight and hearing, colour discrimination and judgements of lengths." [26] Of the 10,000 persons who offered themselves for measurement in the measurement hall, it is recorded that not all were sober.

The exhibit was sufficiently popular that at the conclusion of the Health Exhibition, Galton's operation was removed to the South Kensington Museum where a permanent staff member oversaw continuing the collecting of data (and three-pences) until 1894, the year after the Chicago Fair where Galton's ten years of data formed part of the anthropometric exhibit. Not only were Galton's results on display, but his his schemes of measurement and statistical analysis were continued at Chicago in 1893. American psychologists, with approval of the newly minted American Psychological Association, charged a fee, took measurements, and gave fairgoers a card comparing themselves to the population averages.

Few demonstrations at the Fair would come to so influence American life, and few would have such world-wide influence in the century to come, than the measurement of individual's ability to learn. This was to become a major industry of the twentieth century.

Galton's shadow, both the intellectual and the entrepreneurial, gave force and life to the first attempt in America to gather data on physical and mental abilities of its population. Galton's inquiries extended to the use of fingerprints for identification of criminals, the typing of character (categorizing of personality), eugenicts, testing of 'aptitudes for learning' and 'intelligence', and into the use of photography for the prediction of who might be likely, because of their genetic disposition, to become criminals. In each case, the study of individual differences held promise. The scheme came to be expanded in time to an attempt to find information on the presumed genetic inheritance of all Americans.

Galton, is, and for some decades has been, out of favor. Neither anthropology nor psychology does much to recognize its debt to him as the theoretical and practical expert and pioneer of variation. Can we imagine a social science that fails to measure variation, or that has no descriptive statistics? No more than we can imagine a biology without the concept of evolution. But whereas Darwin's views and life are treated with interest, respect, and awe usually reserved for the reliquaries of saints, Galton and his achievements receive but passing mention in the histories of social science. While Darwin seems to us to be reasonable, painstaking, and kind, Galton rankles the contemporary mind. Galton's views on whom should be educated at state expense, the care of the insane, on the propriety - indeed, the necessity - of class, strikes us as evil; surely, undemocratic. Few readers of his work can separate his achievements form his motives. While Darwin has had thousands of publicists, Galton has had only a handful, most of whom appearing to be more enchanted with his racial and class views than with his grasp of measurement. Before we make judgement, however, we might examine our own culture more carefully than we usually care to do.

Galton's legacy is a wide umbra, one that affects our lives as much or more as the ideas of any thinker. Who among us has not had their future decided or altered by the outcome of a test of mental or physical ability? The idea that variation may be used to predict is an idea of enormous significance to our cultural systems. If the Anthropological and Psychological Exhibits in Chicago can be said to have had a prompter, Galton is a leading candidate. But because of his belief in the inevitability of the dictating power of genes, Galton's shadow would extend far beyond Chicago and the eventual destruction of the Fair. Speaking of which, it is time that we finally wished farewell to the fair, and searched for further pastures.

We Depart

The Fair that has revealed much for us, the World's Columbian Exposition, ended on October 2, 1893 with a grand display of fireworks. Some of the towering, white, Renaissance buildings were put to permanent use, not the least of which being the development of the Field Museum. Most buildings unfortunately became, within six months, winter homes for people and animals needing shelter, 'vagrants' as they were called then, the 'homeless' of today. Eventually their presence annoyed the better off citizens, as it does today, and the authorities decide to reduce the remaining buildings to rubble, thereby scattering the undesirable locals who had taken to living there. These inhabitants were the last human beings to see and to be seen at the Exposition, an ironic continuation and ending to the displays of native peoples and their customs.

Chicago, a phoenix-like city, had displayed itself as having risen from the ashes, and Jackson Park will stand as a sometime verdant, sometimes icy reminder of what was once the world's greatest collection of all humanity has created and designed - and of what people believed about themselves. Most of the folk who formed the living tableaux will return to their lands and cultures, although some will integrate themselves, so to speak, into the life of the city.

The intellectual life promoted by the Fair is not in rubble and is not lost, but transformed. The University of Chicago, built adjourning the fairgrounds, remains a testament to the birth and rebirth that accompanies evolution, whether of human physical structure, or human ideas and institutions. Dwight Mood's evangelical meetings, too, would disappear, but something of his evangelism and teaching would live on in the Bible Institute he founded. Annie Besant would become a leader in the spritist movement, working with and then against Rudolph Steiner, who we will meet in the next chapter. Samuel Gompers would become a major force in the unionization movement in the years ahead. Jastrow would promote the study of individual differences as the basic data of psychology. Woodrow Wilson's idea that physicians and lawyers should be educated first and professionals second would take hold, as the four-year college or university experience became widely seen as necessary to professional studies in many fields, but surprisingly not in education, although Wilson had an agenda beyond his college presidency. The ideas we seek are no longer to be found in Jackson Park so our hunt for mental fossils there is over. Yet these ideas refuse to be buried, they make their way into the public consciousness in ways difficult but rewarding to trace.

Just as the Fair brought together people of all backgrounds, kinds of work, ages, races, genders, sexes; just as it displayed the unusual, the freakish, the rare; so too it brought further into the public conciousness the measure of human variation — measurements so, quite frankly, skimpily, done in the exhibits. This idea evolved into mental testing, intelligence testing, placement in work and education by test-score, decisions about one's gifts and limitations, evidence as to how the races and genders differ. It would give support to attempts to categorize people's mental qualities based on their appearances, and it would give support to the eugenics movement.

But before we see the future of those ideas, we explore the consequence of the fair in terms of the marvel of electricity and the application of photography. At the fair, the first turned night into day by unseen power unleashed, it would seem, by the mere pressing of a button. The second, much featured at the exhibits of anthropology, neurology, and psychology was seen as a technique that allowed for permanent recording of events, people, and human variation. We explore the impact now the power of the mental fossil represented by unseen power.

Chapter 4

Unseen Powers: Spiritism and Other Worlds

In 1850, had you touted your belief that there were rays that could see through flesh, or those that could transmit messages through air, such would have been grouds for considering changing your residence to an asylum. However, the proposition that there was unseeable energy of this nature became a hallmark of modern science shortly thereafter and ever stranger unseen forces have become staples of today's science.

The same proposition was and is a hallmark of spiritist thinking, and also, alas, of schizophrenic thinking. At the fair and in our earlier tentative digs we touched upon the idea of unseen power in several forms: in animal magnetism, hypnosis, electricity, and photography, these often combining with the most influential idea of human thought, that of another world, a heaven or paradise, perhaps an alternative universe. For some, the latter idea is fleshed out with angels interacting with human beings. Some folk attempt to translate between seen and unseen worlds, most often through religious postulates and hypotheses, though sometimes, as we shall read in this chapter, through mathematical formulae. This characteristic of mental fossils I call 'spiritism', a nineteenth century word to be sure, but one that when distinguished from 'spiritualism' suggests an evolved and ongoing attempt to explain human affairs by unseen powers.

One backdrop for modern spiritist views may be said to have been set by Immanuel Kant [1724-1804], anthropologist and philosopher, whose 1797 Critique of Pure Reason is concerned with how 'appearance' can be related to 'reality'. Some would say that all serious modern western thinking about ourselves derives from Kant's understanding of the human mind. Kant argued for our understanding two worlds, the worlds of 'phenomena' and the 'noumena'. The first is the world we know via perceptions, which become organized by the mind into our understanding of reality. The second is what might be called 'ultimate reality', this being composed of the essence or core of things, the 'things in themselves,' as Kant called them; the 'thing' in its most elemental form.

In the current chapter, the idea of unseen power, introduced as electricity at the Fair, is shown to appear in the ideas of Gustav Fechner (1801-1887), Carl Jung (1875-1961), Rudolph Steiner (1861-1925), and Sigmund Freud (1856-1939). [1] All four theorists are concerned with how the unseen is related to the seen, the felt related to the observed, the appearance of things to their reality. Freud and Fechner, whose achievements are often understood to represent antithetical ways of understanding ourselves, both postulated an unseen world as a means of explaining the world of perceptions and human understanding. Their contributions, pioneering work to our understanding of experimental psychology and psychoanalytic thought, are major fossil markers of the 'power of unseen worlds.' Fechner invented mathematical techniques for measuring the relationship between the worlds of 'appearance' and 'reality'. Freud distinguished the conscious from the unconscious, with the unconscious understood to modulate between reality and appearance. Jung too postulated an unconscious that mediated personal experience (the personal unconscious), but added an additional one mediated by evolution and characteristic of all people (the collective unconscious). Steiner, the most practical theorist of the group, held to a belief in unseen power that gave rise to an educational system active today.

Spirit and Healing

A frequent and constant way in which the idea of unseen power makes itself known is through expression of religious faith and through healing of mental or physical ailments. Faith and healing are associated, the one almost always involving the other. The promise of the ability to heal the minds and bodies of the ill is among the most prominent mental fossil evidence of spiritism. We pray, we hope, we have faith in being healed: we hand to others the responsibility of our healing. We expect that some part of us will survive in an existence in another world, even if one now unseen and unknown, perhaps even one worse than the one we now experience. Each of the thinker's ideas who we consider in this chapter had great influence over the thinking of their times (and ours) regarding human understanding of itself. Each has in common hypotheses regarding how and why we human beings react. Compare the statements below, for they apply equally to Mesmer's work with animal magnetism and Freud's with psychoanalytic thought (and perhaps much of modern medicine).

Here is the pattern: The putative healer discovers an unseen, but powerful force or spirit, whose energy governs the state of the mind and body of the human individual; At first, the discoverer is alone able to use the power, but at some point other practitioners are taught and allowed to use the power; The nature of the power often requires some special relationship between the holder and the recipient; In time, the holder of the power transfers it to another person who is able to act as healer; When the power is transferred , the recipient often has an awakening, a realization of a new life or state of being; an understanding of oneself previously denied or hidden; The notion of their being such a power, or at least of the power having the characteristics claimed by the discovery, is attacked by established professions; The process is then practiced by a chosen few followers. Electricity and photography, now understood and established realities, were, at their discovery, examples.

There is a pattern. Victor Race's cure by the Marquis de Puységur. These aspects were evident: a special relationship existed between Victor and the Marquis; the one was master, the other worker; one was seen to have the power to cure, the other wished to be cured. The unseen power that was transferred caused Victor to leave his mind, to enter another mind of another place and time; upon his mind's return to our world, it was changed by being healed. The person in trance appeared to show the memories and perceptions of a person different from him or herself in the normal state. Once released from hypnosis, there was no memory of the person who had appeared during the crises state. The healer could implant in the hypnotized person ideas and instructions that would influence the awake person and be acted upon, yet unremembered.

Dr. Mesmer celebrated his ideas and cures both as a scientist and as a physician. He became a celebrity, a status always dangerous for scholars and scientists, not to mention physicians, clergy, and authors. Patients were eager to see him, while his sympathy and policy of low or no fees for the poor gave him a reputation as a man truly interested in the welfare of his patients and humankind. Mesmer was celebrated for his humane-ness and healing. Mesmer's written works showed decreasing interest in the workings of 'animal magnetism' and the 'fluidum', the force that was transferred between patient and physician and whose stability signaled good health.

It was the curative aspect of the trance that came to hold his interest and to occupy his time. The doctor developed a method of animal magnetism for groups. The utility of group healing may have monetary motivations, of course, but there is an occasional claim that the group itself is a critical aspect of the healing. Mesmer's mass healings required a baquet, this a large wood bucket into which twenty or so people could be contained and arranged. The tub contained iron rods connected to magnetized jars of water. Music was provided, often by Dr. Mesmer himself, who played the glass harmonica invented by Benjamin Franklin, in Paris at the time. Franklin would become the chair of the Royal Commission appointed to assess the validity of Dr. Mesmer's claims of animal magnetism.

Some describers say the patients held hands. [2] In time, many patients fell suddenly into the crises, this being a disengagement of the mind exhibited by shaking, loss of consciousness, and eye-rolling. The crises were curative: those taken by it were removed to a room wherein they rested, recovered, reacquired their conscious selves, and often enough felt themselves healed. A splendid description of the group therapy is provided by Ellenberger: (Note how the several elements of animal magnetism, hypnotism, electricity, healing, landscape, tree-worship, and the postulation of other worlds come together in this taut description from France in 1785.)

"The number of patients became so great that Puységur soon organized a collective treatment. The public square of the small village of Buzancy . . . "was not far from the majestic castle of the Puysegurs. In the center of that square stood a large, beautiful old elm tree, at the foot of which a spring poured forth its limpid waters . . . Ropes were hung in the tree's main branches and around its trunk, and the patients wound ends of the rope around the ailing parts of their bodies. The operation started with the patients' forming a chain, holding one another by the thumbs. They began to feel the fluid circulate among them to varying degrees. After a while, the master ordered the chain to be broken and the patients to rub their hands. He then chose a few of them and, touching them with his or on rod, put them into 'perfect crises.' These subjects, now called physicians, diagnosed diseases and prescribed treatment. To 'disenchant' them [that is, to wake them from their magnetic sleep], Puységur ordered them to kiss the tree, whereupon they awoke, remembering nothing of what had happened. [3]. . . It was reported that within little more than one month, 62 of the 300 patients had been cured of various ailments." [4]

Rudolph Steiner and the Anthroposophists

Rudolph Steiner pursued a life-long attempt to construct a philosophic, educational, and medical system that seems to have no need for hidden messages, rapping from the dead, or spirits who speak only through designated living humans. [5] If the experimentalists wanted to organize the material world of our perceptions and realities, Steiner wanted to reveal the spiritual energies that guide the individual and the planet. His system came to be called anthroposophy, meaning the 'wisdom of mankind'.

Steiner's position was that the spiritual world could be perceived most clearly through utilizing the human capacities for intuition and feeling. His position reflects a theme that is long-standing, but one that, in our times, is generally devalued to favor the human capacities for thinking and reasoning. We live in times during which materialism is dominant; when reductive science is used to filter our understanding of our universe and ourselves. The person who understands the world only through materialism is, goes the oft cited simile, only a sleeping man, one who has a few visions but knows not what to do with them - one unable to find his or her location within the universe and the causes of her or his actions. The study and practice of anthroposophy allows one to refine the imagination, to understand the essences of perceptions, to distinguish the essence from the perception. As a metaphor, let us realize that a book can be perceived, as can the letters that comprise it, but once read, the book has an essence and provides a totally different perception from the nature of the book merely seen as a physical object. This observation is the metaphorical core of Steiner's educational system.

Both Steiner and, earlier, Emanuel Swedenborg's (1688-1772) views prompted an educational system that remains active. Schools based on Steiner's understanding of the spiritual development of the human being are active in the UK, Germany, The Netherlands, and in the United States in a number of cities. Known generally as 'Waldorf schools', so named after a major financial donor, in these schools human imagination, a necessity for locating the spiritual world, is promoted and developed through rhythm, music and singing, the repeating of sagas, legends, ancient history and mythology, and stories from the Old Testament." (6)

Rudolf Steiner offered a set of five lectures to 1,700 persons at an educational conference of the Anthroposophical Society of the Free Waldorf School in Stuttgart in April of 1924. They were nearly the last public lectures he gave as he became ill and died a few months later. He took as his main theme the question "What is the position of education in the personal life of man and the civilization of the present day?" [7] What strikes this contemporary reader first is the attention given by Steiner to the need for the teacher to teach his or her ownself so as to create the symbiotic relationship necessary to transfer information, ideas, and intuition to the child. The accent on the teacher training his or herself to be intuitive and imaginative is surely different from the standard view that a teacher's task is to transport information to the child through an understanding of how people learn. Without data, I suspect that the training of teachers today focuses on how the teacher is to transmit information that can be tested; but, writes Steiner, this is wrongheaded: teacher and student alike can look up the capitals of countries, correct spellings, or work through formulae. To compel rote-learning of these matters is to mistake a book for its contents. To Steiner, the teacher's role is that of facilitator, for all education must be self-education. The true teacher is an active facilitator, for the teacher has nothing to teach unless he or she is actively engaging his or her intuitions and feelings.

A compelling difference between Steiner's recommendations and common practice today is the emphasis on the personality of the teacher. Consider this revealing (and typical) passage by Steiner in regard to the temperament of the teacher:

". . . let us take our start from one of the qualities of man's nature, the temperament. . . we will not in the first place consider the child's temperament — for there we have no choice, we shall have human beings of every kind of temperament to educate — but the temperament of the teacher. The teacher enters the school and stands before the child with a definite temperament that may be either choleric, phlegmatic, sanguine or melancholic. The question; "What ought we as educationalists to do in the way of a possible curbing or self-discipline of our own temperament?" can only be answered if we bear in mind this other fundamental question: "What effect has the temperament of the teacher, simply because he is there, on the child?

"We will take, firstly, the choleric temperament . . . He may do all kinds of things in the environment of the child which frightens it — for we shall see how delicately the child's soul works. The fright may pass quickly away, but none the less be implanted right down into the physical organism of the child . . . or it may be that the child always has a sense of terror when it comes near the choleric teacher or it may feel a sense of oppression . . . Take a child who is still at an early age, before the Elementary School age. It is then a wholly uniform being. The three members of man's nature, body, soul and spirit, separate from each other only in later life.

"The effect of the phlegmatic teacher on the child will be to leave its innate activity unsatisfied. The inner impulses want to come out, indeed they stream out: the child wants to be active. The teacher . . . does not respond to what comes from the child and thus what is striving to flow out finds no impressions or impulses to meet it. . . .The soul of the child feels a kind of suffocation if the teacher is phlegmatic". [8]

Three further issues dominate Steiner's thinking about education: each reveals an aspect of his understanding of spiritism. First, natural science is not to be derided: it produces valuable and useful information. The danger to the individual's body, soul, and spirit, occurs when he or she comes to believe that there is no verifiable world other than that discovered by natural science. He finds that human (read 'European') history reflects a shift away from interest in the spiritistic world. Aristotle and Aquinas well understood the importance of identifying and understanding the essences of things and the medieval church concentrated on the spirit and largely ignored the physical and materialistic. The period we know as the Enlightenment, paradoxically, produced a civilization that not only favored discoveries solely within the material world, but came to distrust reflection on the spiritual worlds, a fact that is seen clearly in how we educate children and ourselves today. One is reminded of Carl Jung's rhetorical question about children's education: But, who will educate the heart?" Education is an art: like the artist working with clay or color, the teacher works to create a novel form from each human being. Knowing merely the name or shape of another human being does not tell us of or about that person: it is through the feelings and unconscious perceptions that we gain knowledge of another human being. The purpose of education is to shape and aid such discriminations, for they are the source of true knowledge. The pattern he sees in (European) history, he writes, is repeated in the development of the individual. The shift from spiritist thought to materialist thought is an example.

The understanding of the infant and young child can be compared to a sense organ, like the eye or ear, something receptive to all that the environment throws at it. Spirit and soul merely imitate impressions that impinge upon the child's world. With the loss of baby-teeth, Steiner points out, the mind of the infant is replaced by that of a new person, one "who develops under the influence of those forces which the human brings with him from his pre-earthly life . . . the forces of heredity belonging to man's physical stream of evolution are fighting with the forces brought down by the individuality of each human being as the results of his own previous earthly life from pre-earth existence." [9]

Humankind comes into this world with mental predispositions, these having been acquired in pre-earthly lives and conditions, says Steiner. The idea resurfaces as the 'collective unconscious' in Jung's thinking. One purpose of education is to attend to these ideas, whatever they may be, and to mold them into an appreciation of the good, the just, and the beautiful. Hence children are interested in, and can be educated by, fantasy and the magical. Steiner suggests that each human being has four kinds of 'bodies': the physical, etheric, astral, and one of ego organization. The etheric [10] has inner-rhythms. These may be trained by the educational system by an activity that appears to others a group dancing, but which is a way of training to assist one to find the inner voice of inner rhythms. The astral has ceased to be material, and is therefore of a somewhat higher order, while ego-organization (not to be confused with the Freudian word ego, the "I am") involves the highest form of integration with the spirits of the other world. Human and animals alike are subject to spiritual energies, hence Steiner's disapproval of meat-eating, not because of cruelty to those eaten, but because by meat-eating one takes in the energy and vibrations of other forms, thereby giving up some of one's own.

When applied to health and healing, anthroposophistry was led by the knowledge of Ita Wegman M.D. She was a physician and anthroposophist, a native German writer skilled in the use of the English language. (Steiner came to suspect that in a former life she was Sabina and he was Gilgamesh, an observation that did nothing to convince the public of the value of anthroposophistry.) Wegman's short manuscript *Fundamentals of Therapy, an Extension of the Art of Healing through Spiritual Knowledge* had the honor of being among the last documents that Steiner read. The manuscript was published, with Steiner as first author, a year after his death and became seen as his major statement about medicine, health, and spiritism. The opening sections speak directly of the postulates that guide anthroposophy:

"If in a moment's introspection we consider our everyday activity of thought, we find that the thoughts are pale and shadow-like beside the impressions that our senses give us . . . A new world begins to dawn for the man who has thus enhanced the force of his perceptive faculty. He, who till now was only able to perceive in the world of the senses, learns to perceive in this new world; and as he does so he discovers that all the Laws of Nature, known to him before, hold good only in the physical world only. It is of the essence of the world he has now entered, that its laws are different, nay, the very opposite of the physical world. Another force emerges, working not from the center of the Earth outward but inversely. Its direction is from the circumference of the Universe towards the centre of the Earth. The faculty of man to perceive in this world is attainable as it is by exercise and training, is called in Anthroposophy the 'Imaginative' faculty of knowledge . . . The word is used because the content of consciousness is filled with living pictures, instead of the mere shadows of thought. [It is called the etheric.] The etheric world of the individual may be understood by thinking about plants. How can we account for the growth of a plant from a seed? We can describe the chemical and physical processes that occur, but in no essential way do we understand how seeds become plants. Forces act on the human seed and embryo providing, in time, a grown person who can be capable of understanding the forces that create the humanly image. Thus the formative or plastic force, appearing from the one side in the soul-content of our thought, is revealed to the 'imaginative' spiritual vision from the other side as an etheric-spiritual reality" [11]

To Steiner and Wegman, many kinds of illness come about when we fail to match materialistic concerns with our feelings. People become ill when they fail to understand their place in nature. Eurythmy, the making of motions, rather like group-dancing, serves to place one in sympathy with their nature: it is an essential aspect of the educational and personal program. Done individually, or as a group, the resulting hand and arm motions are like poetry and are regarded as a form of speech. Indeed, Steiner proposes the motions parallel the evolution of speech in human beings. The reaching toward naturalness assists diseased organs. The nature of food eaten and chemicals consumed are also responsible for health. The body is continually disintegrating and reintegrating, as it is subject to the forces of the universe. Consider acute pain from 'stoppages' or 'congestion' in the lower abdominal area. "We can observe, in the cute attacks of pain, an excessive activity of the astral body" [12] Let us consider an abstracted case and its remedy:

"A man of forty-eight years. He had been a robust child with a healthy and vigorous inner life. He had undergone a five-month treatment for nephritis and been cured. Married at 35, he had five healthy children. At the age of thirty-three he began to suffer from depression, weariness, and apathy. He began to feel helpless mentally and spiritually. He is confronted by endless questions and misgivings in which his profession [schoolmaster] appears to him in a negative light. This pathological condition reveals an astral body which has too little affinity with the etheric and physical and in its own nature is immobile. The physical and etheric bodies are thus enabled to assert their own inherent qualities. The feeling of the etheric not being rightly united with the astral body gives rise to the depression, while the deficient union with the physical produces fatigue and apathy. The whole condition indicates that it is necessary first to strengthen the astral body in its activity — a thing that can always be attained by giving arsenic in the form of a mineral water which [along with highly diluted doses of phosphorus] supports ego-organization. An Eurthmy treatment re-establishes the harmony. We had the satisfaction of observing a complete cure in this case." [13]

The educational and medicinal beliefs of anthroposophistry probably have aspects and elements that speak directly to each of us and our evaluations of our own wants and needs. As was true of Jung and Freud, but not Mesmer and Fechner, Steiner offers us a comparatively complete analysis of our behavior based on the principle idea that other worlds, along with previous worlds, influence our minds and bodies. Some of his ideas we consider to be modern (these we call 'insightful'). Others we find to be medieval (these we think of as pre-scientific). Whichever, Steiner's system is easily as comprehensive as any contemporary theory of the mind and its history.

It is difficult, perhaps impossible, for us to put aside the materialism of the folk psychology of our time. At the very least, nonetheless, we owe it our quest to place Steiner's ideas within the record of mental fossilts of understanding of te mind that accept power and presence of alternative universes. Steiner, by the way, was no stranger to experimental psychology: he received a Ph.D. approved by von Hartman, one of the leading psychophysicists of his time, himself a former student of Gustav Fechner. In academic lineage, Steiner was a direct, second-generation descendent of the beginnings of experimental psychology. If it seems odd to us now that Steiner's views would come out of such a backround, perhaps we need to dig a bit earlier and uncover the beliefs of Fechner himself.

Emanuel Swedenborg

Franz Anton Mesmer

Gustav Theodor Fechner

Amand-Marie-Jacques de Chastenet

Rudolf Steiner

Annie Besant

Figure 4.1: Theorists whose work illustrate the concept of 'unseen worlds' of power; forerunners of the notion of unconscious thought. Kant's argument that human thought about the material world is necessarily constrained by our categories of perception was denied by these theorists who suggested examination and development of the internal, spiritual world of consciousness.

Douglas Keith Candland

Fechner invents psychophysics to compare the unseen and seen worlds

Experimental psychology, focusing as it does on the empirical and the material, would appear to be unwelcoming to the idea of spiritism, I suggest that experimental psychology is rooted in the spiritism of European thinking of the 19th century.

Gustav Fechner's (1801-1887) psychophysics, which formed the keystone of nineteenth century psychology, was an attempt to measure the world of spirits and psychic energy. As psychology 'matured', talking to much of this was seen as embarrassing and the idea was disregarded, ignored, eventually forgot by academic psychology. Fechner lived through the span of the nineteenth century, a time when, in the West, German was the language of culture, science, and scholarship, when Americans who wanted to learn the latest techniques and the newest ideas found it necessary to acquire time for study, and, preferably, an academic degree from the German universities and laboratories. Many who wanted to study the new psychology, the experimental sort rather than the animal magnetism- and hypnosis-sort, studied with Fechner's student, Wilhelm Wundt, [1832-1920] at the University at Leipzig, for here the new experimental science involving the measurement of human experience and human reaction was practiced, taught, and brought to fruition with experimental publications in the new journals of experimental psychology. [14]

Trained in medicine, then a course in general science rather than in the practice of medicine itself, Fechner taught physics, then physiology, and, according to later historians, invented experimental psychology by the 1860 publication of Elemente der Psychophysic. There was another side to his personality, for when not doing or teaching science, he wrote under the psyudonym 'Dr. Mises', and published scientific and religious satire, works on esthetics, and argued for a cosmological system that was in tune with the German intellectual tradition of the time, seeking to encompass all imaginable worlds under a single principle. His speculations late in his life on other worlds can be understood was a continuation of the problem addressed in the *Elements of Psychophysics*.

A serious illness preceded Fechner's shift in research: G. Stanley Hall, celebrated American psychologist and university president, who knew Fechner, if much later than the event he here described, tells us that during the illness Fechner "could not control his attention, and his thoughts were very active against his will. He could not sleep, and became depressed and tense by turns. Some optical disease set in, and he was obliged to give up all reading and writing, and spend much time in a darkened room. Even dark glasses were intolerable, and he found his way about the streets almost like a blind man". [15] The episode is not restricted to Fechner's life: each of the other people whose work is described briefly in this chapter on spiritism and psychology, Mesmer, Freud, and Jung, suffered like crises, usually in early middle-age. Hall himself was to loose his eye-sight. The similarity of these crises to the crises experienced by Mesmer's patients should not escape us.

While he recovered from his illness, what took Fechner's attention was how one might move back and forth between the world we sense and the world we imagine, between the worlds of appearance and reality, as

Kant put it. Fechner reasoned that just as the world of our perceptions is lawful, so must be the other world. If both are lawful within themselves, then it should be possible to show that they are lawful in relation to one another — if only we could learn how to measure the lawfulness of one in respect to the other. Fechner tells us that on the morning of October 22, 1850, in bed, awaiting the serving of his breakfast, he understood how solution.

The methodology was seemingly simple; the mathematics not. Ask subjects to compare perceptions, to judge their difference or alikeness, or sometimes their presence or absence. Then, mathematical and statistical procedures could be used to determine the laws that relate our perceptual world to our material world. Evevtually, Fechner proposed that, relative to the physical world, the psychical world changed logarithmically. If a 100 watt light must be increased in brightness by 10 watts in order for me to judge it as brighter; then a 1000 watt light will need an increase of 100 watts. The difference, materially, is either 10 or 100 watts, but the difference perceptually is identical, this being one unit of 'change.' [16] By these methods, psychophysics would show that parts of the mind could be measured independently, and, once measured, could be shown to relate lawfully and mathematically to one another.

The resulting *Elements of Psychophysics* was to be for experimental psychology what Darwin's book of the same time was to natural history. The 'psychophysical law' was not revealed sparingly, as was, say, the structure of DNA strands. Although running to 585 pages, the 'fundamental formula' given above is established in the first two chapters; the rest is documentation. A chief concern of Fechner's was to explain that the real and physical worlds are, in fact, merely reverse sides of the same world, the one side observable, the other, much less so. This, however, left Fechner with many unanswered questions: Where, then, is consciousness? Is it to be explained at the level of the atom; molecule? cell? assembly of cells? Is consciousness, then, another matter for materialism to uncover? Or is it something else; something more? Some thing that only a belief in spiritism can reveal? Fechner saw his Psychophysical Law as demonstrating the parallelism between these worlds, the "Psychophysiche Parallelismus im Universum", as he called the problem.

The indefatigable Fechner wrote also on issues of the day. The nature of these issues tells us something about the origins of social science at this time and place. They included whether plants are living and conscious, the likelihood of their being atoms, and the likely path of the evolution of consciousnss. We find in the writings of Fechner/Mises an interest in the soul, by which is meant a spirit existing in infinite time of which our body on this earch is but a passing phase. Fechner became interested in understanding what the dead are doing. They were being active, to be sure, and influenced the course of events on Earth. They speak to one another through us, a fact that leads us to misunderstand our own, and other's, motives. When we think of the dead, we thereby give them life. We create by our thoughts a world of spirits, but we have no way of knowing the reality or modes of functioning of these other worlds. Like Kant, Fechner places the creation of reality within the human mind, but unlike Kant, Fechner seems to want the other worlds to be both 'other -worldly and something with their own independent reality.

Douglas Keith Candland

Steiner and Fechner came to their different ways of answering a like problem: how can the worlds of reality and appearance be translated, one into the other? Late in his life, Fechner published two books on the world of the spirits, the Zend-Avesta (from the Parsi, 'spirit world') and the Äesthetic. The first was a summing up of his spiritual and religious beliefs, unencumbered by considerations regarding the empirical or sensory world. There is one God (Fechner wrote) all living things - animals, plants, and people are given life by the same energy. Above them is Jesus. Humanfolk are not born or die in the sense of the spirit's doing so; it is only the material aspect that does so. In this life, the task is to advance The Creator's design and to learn to judge our feelings. This life is but "a seed in the cold ground, awaiting the growth and development that comes through the nourishment of heaven." Doing good and evil is not always judged in this world, but surely in the other world. This book was read, and admired by William James, who, in recommending it said of Fechner "He seems to me of the real race of prophets. His day, I fancy, is yet to come." [17] The second book takes up the esthetics of beauty, wherein Fechner emphasizes the experimental treatment of humankind's perceptions of beauty. The approach is, thereby, somewhat statistical and utilitarian, an approach very different from the esthetics proposed by Plato or Kant.

Examining Fechner's contributions is akin to examining the facets or aspects of a gem: whichever way we turn it, different shapes and colors are reflected and revealed. At Fechner's funeral in November, 1887, Wilhelm Wundt honored Fechner for his attempt to understand human existence by connecting humankind's spiritual and materialistic sides. This provides rather clear evidence of how Fechner was understood in his day. It is ironic that contemporary psychology honors Fechner not for this, but for his experimental materialism, failing to remember that the impetus for the experimental work came from a desire to grasp the relationship between the real and the spirit worlds.

Sigmund Freud and the Spiritist Tradition

We prospectors of mental fossils can find similarities between the lives of Mesmer and Freud although the lives were spent two hundred years apart. Both were trained as physicians and both developed a world-view of what held the universe of behavior together. For Mesmer, it was the 'fluidum', for Freud the interplay of the conscious and unconscious. Both hit upon their novel methods of healing during their middle -age years; both had a period of personal withdrawal and self-doubt followed by an energetic and productive period; both were attacked by the medical establishment of the day; both spent the last part of their lives defending the system of healing, developing it, and dealing with their often quarrelsome followers; each knew both sides of celebrity-status, the honor and the criticisms; and both, for very, very different reasons, and in different ways, spent their last years in something akin to exile.

Freud tried to find an unseen power's effects on the body, arguing that that the 'other world' was to be found within, not outside, the human mind. While Mesmer looked for the invisible force in the air, Freud found it in the mind itself. His idea was a logical extension of the discovery of other worlds existing within the individual, as revealed by hypnotic states. The mind, Freud claims is not composed of separate faculties such as reason, feelings, emotions, and motivations, each complete and set along side one another, as was thought at the time. We must rotate our perception of the mind, says the revolutionary idea, to see that the mind is, in fact, a single entity composed of three faculties mingling and working at once together, yet constraining one another. The model of electricity, visited at the Chicago Fair, is re-lighted. The power of impulses comes from the id (the Es), but like electricity, as understood at the same, it is an useen power. The core of the id has properties much like those projected for the source of animal magnetism, because they are filtered by the superego (the Uberberich) they become unknowable, the world of the id forever unseeable by us. (for electricity, read resistance and output). The interaction yields the ego (the Ich), our way of dealing with the material world, composed of ideas, symbols, substances, and categories on which to hang our perceptions of ourselves and those around us. [18]

As Freud's ideas can be seen as a continuing development in humankind's wish to understand its own minds, his intellectual relation to Mesmer shows a somewhat clearer line of development than is usually appreciated. Pioneer followers of Freud, in their urge to promote his metaphor of the mind through psychoanalysis and to thereby revolutionize our understanding of the mind, became, as was true of Mesmer's followers, highly protective of the intellectual leader's ideas to the point that new ideas were thwarted. To promote psychoanalysis, societies were formed in Europe and America (Berlin and New York Psychoanalytic Societies), at first organized for the presenting of ideas, then, increasingly, as feudal societies intent on keeping out the ideas of those without 'standards'. In America, this form of intellectual protectionism led to psychoanalysis becoming a medical specialty, with all the protection under the laws of healing and medical practice that such involves.

On the first brush of this canvas, it appears that Freud and the other worlds of spiritism have nothing in common. Freud's is a hydraulic system, perhaps a pneumatic or electrical system, but not obviously a spiritist system, although David Bakan has presented the idea that Freud's theory has a debt to Jewish mystical thinking, perhaps to Kabbalistic ideas. [19] There is a spiritist component to Freud's system, in so far as we are willing to consider the ability of the mind to substitute symbols for the material as akin to spiritist notions of other worlds. Every dream has both a manifest (real) and a latent (symbolic or spirit world). When Freud unmasks dreams by deducing the latent content from the manifest, he is doing what Fechner did, a sort of psychophysics, for he is finding the psychological Rosetta stone that allows translation among, and between, the real and the unseen. Slips of the tongue, mechanisms of defense, and other standard Freudian ideas can be understood as ways in which the material world is translated into the spiritist world, and back again.

Freud pointed out, in his later years, that he had been born before (1859) The Origin of Species was published. He was, in lifespan, both a pre- and post-Darwinian [20]. Freud's central work respecting the ideas of evolution is to be found in Totem and Taboo, a work that many of Freud's followers dismiss, for reasons themselves deserving of analysis. Contemporary interpreters of psychoanalysis mostly ignore Freud's interest in evolutionary anthropology, rather as contemporary psychologists disavow Fechtner's spiritism. Freud was a committed Lamarckian, to the point of rejecting the advice of his colleagues to consider the new discoveries in genetics of recessive and dominant characteristics; however, Freud had a serious interest in how humankind developed and evolved its beliefs. The Oedipal complex, for example, was seen by Freud to have evolved in the thinking of humankind, just as physical structure was understood to evolve. Before positioning himself as a student and theorist of the mind, Freud published researches on anatomy and physiology, including work on the acoustic nerve and the testicles of eels. He was, thereby, more than alert to the Darwinian approach that, although slow to achieve acceptance in Germany and Austria, was nonetheless well-known among scientists.

Totem and Taboo offers a direct attempt to provide a psychoanalytic theory of the mental evolution of humankind. The contemporary origins of the idea are to be found in Darwin's *Descent of Man*, but the overlay of psychoanalytic thinking on to the proposed anthropological history of human evolution provides a scenario in which early man lived in male hordes under the dictatorship of an elder, male leader who controlled young males and all females, conserving for himself resources including food, water, and available mates. Such an arrangement, although neither Freud nor Darwin commented on this, is known today in the societal arrangement of certain nonhuman primates, especially the baboons. [21] Freud developed and embellished the scenario by imagining that at some point the young males banded in rebellion. When successful, the revolution resulted in the ouster of the alpha male or, in eating him, symbolically or otherwise. The new leaders also expelled the females (to whom he might be related) and mated with previously unavailable females captured or found wandering from other troops. In fact, something like this is the way in which some living nonhuman primates manage to alter the gene pool while living in a society dominated by a single older male. Freud speculated that from this primitive scene, contemporary human beings retain a sense of guilt, as in 'original sin', the ideas of redemption, sacrifice, restrictions on sexual relations with near relatives, and the like. Through these speculations, Freud attempted to provide a history of the evolution of certain ideas identified conceptually with psychoanalysis, such as the Oedipal complex. While Darwin's postulations described how behavior, especially mate selection, was needed to enhance variability if the species was to both continue and yet change under the force of natural selection, Freud's backwards projection added descriptions of the evolution of the accompanying mental states.

Jung and the Spiritist Tradition

Going beyond Freud's tentative offerings, Carl Jung's postulation of the 'collective unconscious' firmly combines the evolutional to the spiritist, forming the audacious idea that aspects of the unconscious are inherited. Perhaps no other single idea so well expresses the application of evolution to the human mind; few have been so derided. Jung postulated the unconscious to have two aspects: the 'personal unconscious', in which one hides or stores one's own experiences and the 'collective unconscious', this composed of the symbols and representations pre-wired into the brain by inheritance. (The relation to Kant's a prior and a posteriori distinction seems overlooked.)

Jung believed the mind and its ideas to evolve, presumably in much the same way as structure evolves, as noted by Darwin in his statement: "It is, therefore, highly probable that with mankind the intellectual faculties have been mainly and gradually perfected through 'natural selection'." This statement is clear: mental and physical evolution occur according to the same principle, natural selection. What Darwin does not discuss, although he had hoped to do so, was how mental and physical evolution might alter one another. To choose, almost at random, from the many such explanations and examples Darwin presents:

"We shall find this view of the force of habit strikingly confirmed ...it will be shown that the hair of the insane is affected in an extraordinary manner, owing to their repeated accesses to fury and terror." [22] and

"We have all of us, as infants, repeatedly contracted our orbisular, corrugator, and pyramidal muscles, in order to protect our eyes whilst screaming; our progenitors before us have done the same during many generations . . . " [23]

Jung argued that we filter our knowledge through 'archetypes'. These were 'perfect forms' or 'essences', similar to Plato's' but not limited to physicals such as 'circles', 'colors', or 'chair'; they including ideas such as, 'mother', 'rebirth', and 'hero'. The latter, of course, are far more complex than the former, or they seem so to our conscious selves. Just as I cannot draw a perfect circle, but yet know an approximation of one when I see or make it, so consider the archetype of re-birth: I can 'approach' it, although not reach and touch it directly, through its symbols; e.g., the season of spring, the female birth canal, baptism. These archetypes recur in literature, religions, architecture, and, of course, dreams. We may not escape them because they are the filters through which we understand our perceptions. Unfortunately, Jung's theory remains ungrounded by evolutional theory, even though it is, in fact, very much an evolutional theory. At best, it pays lip-service to evolution. Merely using the word 'evolution,' however, is not an act of explanation. If we explain the supposed racial and cultural differences in symbols and archetypes by recourse to something we call evolution, then we must also do the grinding observations that evolutional theory demands: we must measure the variation and show how and where natural selection is at work.

The Evolving Idea of Spiritism

When we unearth and explore the theories of thinkers as seemingly diverse as Mesmer, Steiner, Fechner, Freud, and Jung, we find a similarity that transcends the appearance it shows temporarily in a particular time and place. We can see, for example, that the idea of a material and a spiritist world dresses and redresses itself, appearing as the 'unconscious' to one thinker and as the 'fluidum' to another. If we are interested in tracing the evolution of ideas, we must have a wary eye for the characteristic that is independent of the particular clothing of a time and place. We can find a suggestion that our mind grasps the connection among seemingly disparate ideas only gingerly and suspiciously. But this is to be expected, as it can be equally hard to see the underlying similarities behind disparate species, for example the closely related, and very similar, wild boar and a hippopotamus, or the mouse and whale (as opposed to fish and whale). Yet, in a seemingly contradictory manner, after our mind identifies such an evolving idea, its obviousness may lead us to think of the discovery as trivial.

The most conspicuous examples of how the human mind seeks to live with in two difference worlds, materialism and spiritism, appears in notions perhaps too generously described as 'religions'. Religions are cultural as well as spiritual, culture being one major way in which the human mind seeks to unite the unseen with the knowable. Among the aspects of culture that influence evolution are cultural restraints and encouragement of marriage and reproduction that favors like-minded religions, thereby influencing mightely the gene pool and, thereby, the evolution of human beings.

Chapter 5

Phrenology as a Mental Fossil

The idea that the structure of our bodies, whether lean, fat, thin, muscled, speedy, red-faced, or pale-faced, tells us something about the traits that govern our temperament, character, and abilities is known to us from classic Greek times. It has been expressed in medical science and literature alike for two thousand years, making it among the most continuously prominent and supect of mental fossils.

The idea exposed itself most conspicuously during the 19th century, especially in the English-speaking world. One proponent we have already met was Francis Galton. In one demonstration, which captures many of the themes to be developed later, he re-exposed photographs of various people ('morphing' in modern terms) with the expectation of deriving and showing the model or exemplary type of the intelligent, virtuous, and talented individual. Further, he superimposed noses to a single image to demonstrate that different races have noses that maintain a common distinguishing property. [1] His aim was to show that not only do people of different cultural and genetic origins have differently shaped noses: it was to show that the shape of the nose could be used to predict characteristics, including criminality.

Phrenology commanded as much respect and utility in its time as mental tests do in ours. Writers Walt Whitman and E. A. Poe used or approved of it. Darwin was invited on the long voyage during which he collected the information that would become his statement of evolution because of his phrenology (the ship's captain being an adherent). The technique was used in prisons and schools to diagnose and label people; to determine what, and whether, they might be taught. Horace Mann, the pioneering educator and promoter of free public schooling in the US, used the technique in sorting pupils, as did Samuel Gridley Howe when establishing an asylum and school for the retarded, and jurist Charles Sumner for judging character. On goes the list of prominent nineteenth century Americans who trusted phrenology and who used it to augment and implement policy in education, criminology, and medicine, in the hiring of college professors.

Today's view of phrenology is that it was prescientific, a silly idea and fad. The idea, however silly it may now seem, nonetheless, provides an important set of fossil specimen in our collection, and we may indeed, be forced to somewhat change our views. The core idea of phrenology remains a powerful way in which we understand one another and can be seen to dictate scientific conclusions. At its beginnings in the early nineteenth century, 'phrenologists' showed interest in all physical aspects of the human body, but especially the size and shape of the internal organs. Only later did the skull become of greatest importance. Eventually the skull became the only 'organ' examined and measured, no doubt for practical reasons for it is easier to observe and measure the head than the liver. Modern ideas reflect the same beliefs that once served as the core of phrenology: namely, that the brain has 'centers' responsible for specific emotions and acts.

The mental fossil-record of phrenological-thought shows that phrenology was originally intended to advance questions regarding the nature of the mind, and became extended to our mind's coming to know the mind of God. The 'faculties of the mind' proposed by 1825 had became fixed when the idea crossed the Atlantic and moved from an academic to a commercial form and setting. The pattern "academic to commercial" appears to be a not uncommon path for ideas that promise to predict human talents and behavior, as Mesmer and others discovered.

Gall's Skull

Phrenology found fertile grounds first in Scotland and England in the 1820s and 1830s, soon to be followed by France, and, less strikingly, in Germany and Switzerland. Like Freudianism and other forms of Spritism, as noted in Chapter 3, phrenology found most prominence in the United States. Its primary scientific claim was that the human mind was composed of separable faculties: examples are parental love, friendship, combativeness, self-esteem, hope, benevolence, mirth, love of colors, and order. Books of the time list between thirty-seven and fifty of these traits. The pioneer phrenologists further theorized that these faculties of the mind were localized within specific parts of the brain; that is, the trait for, say, 'destructiveness' is located above the left ear (in the temporal lobe, we moderns would say), 'parental love' at the lower rear of the head (our visual cortex), and our sense of musical tune over the eyes (exactly where we today think this ability to be stored).

Both of these beliefs dovetail well with modern understanding. Any number of measures of personality in frequent use today postulate faculties of the mind, kinds of behavior, or personality traits, such as intelligence, aggressiveness, and neuroticism. We too believe that separate parts of the brain serve different kinds of perceptions, emotions, and behavior. These two hypotheses basic to classical phrenology continue to be accepted in our times. Alas for scientific method, the validity of each is tested only by its correlation with the other concept. As neither is thereby independently assessed, what looks to be a scientific finding is in fact merely a semantic tautology. Modern mental-testers understand that correlation is not causality, but in practice such knowledge is often not honored.

Skull- oriented phrenology required one to formulate a further postulate: that the areas storing the faculties can be identified by size and shape, and that irregularities of the skull's surface reflect the development, or the lack, of the (literally) underlying faculty. The first half of the postulate forms the core of modern neurology, the latter is seen as uncertain or dismissed outright.

Let us consider how the ideas of the founders of phrenology became altered for by so doing so we might find evidence of how mental fossils evolve. Franz Gall [1758-1828] and Rudolph Spurzheim [1776-1832] Gall's collaborator, understood themselves to be anthropologists, a field of study more or less invented by Immanuel Kant through the course with that title he gave at Koningsburg. Together, Gall and Spurzheim toured Europe and England, giving lectures and demonstrations both at universities and as part of the cultural diet of towns. They were, by their method of spreading news of the technique, the celebrated savants of their day, kin to those today who use radio, TV, touring, and interviews to push forward their pet discovery and ideas. While touring in Germany and Switzerland, the two had a disagreement that resulted in their going separate ways. (In similar fashion, Freud and Jung's visit to the United States was to lead to a rift.) Spurzheim visited the United States in 1832, lecturing in Boston, Yale, and Harvard – where he died. During the tour, he investigated the criminal mind by measuring the skulls of those inhabiting the Massachusetts State Prison.

In what was to be his last book, Gall (1822) summarized the status of phrenology, or what he called the psycho-physiology of the mind, the term 'phrenology' not having his preferred name for the study. Here are the summary points:

1. Moral and intellectual qualities are innate,
2. Their functioning depends on organic supports,
3 The brain is the organ of all faculties, of all tendencies, of all feelings, [very modern, this emphasis; or very aged,],
4. The brain is composed of as many organs as there are faculties, tendencies, and feelings.[2]

Gall provides us with a description (written in third person) of his original finding of the location of 'amativeness', this being a term he coined for the love-center of the brain: Gall discovered it early, by accident, in a young widow patient who was the victim of 'periodical nymphomania,' by often observing, while holding the back of her head in his open hand, that it was both very thick at the nape of her neck, and very hot, and drawn back by its natural language, while she was suffering from its paroxysms. His knowledge of her inordinate passion, along with this thickness and heat, suggested the existence and location of this faculty and organ. [3]

As suggested om the technique described and conclusion reached, Gall's own skull must have showed extreme 'amativeness'. Ackernecht and Vallois, who have produced the fullest account of Galls life and a catalog of Gall's collection of skulls and casts, write:

"Of his three passions: science, gardening, and especially the third one, women, none was very conducive to create a peaceful home atmosphere. His letters are full of allusions to different mistresses and to an illegitimate son, Hamann. Once, Gall declared candidly: "Neither sin nor friends will ever leave me." [4]

By the 1870s, Ferrier in England, and Fritz and Hitzig in Germany, had shown that the cortex of the monkey brain, when stimulated with small intensities of electrical current, responded by initiating motor actions specific to the area stimulated. Electricity was again showing itself, as unseen as it was, to have dramatic power. Not only did it power lights, but also the brain and behavior. Stimulate this part of the brain, and the finger moved; there, and the toes moved. Ferrier's drawing of localizations (Figure 5.1) could be reprinted in a modern textbook without much change.

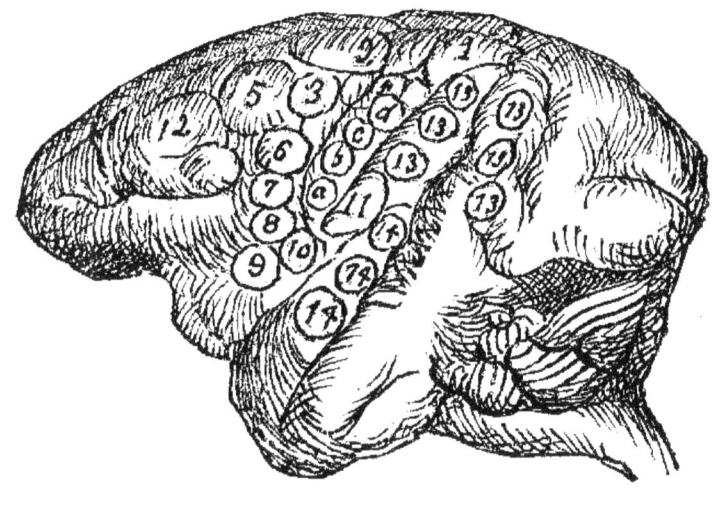

MOTOR CENTRES IN MONKEY'S BRAIN.

Figure 5.1: 'Motor Centres in the Monkey's Brain'. From: Wells, 1898, p.46). The locating of such 'centers' gave plausibility to the basic idea of phrenology that specific areas were responsible for specific traits. In neurology and neuropsychology, the search for centers continues.

George Combe and the Constitution of Man

What determines the nature of the faculties we investigate? How is it decided how many kinds there are, what they represent, the nature of the concepts that describe mental and physical traits? In short, what is the source of the categories? Why "Secretiveness" then and "Perception" now? Why "Benevolence" in phrenology and not "Mesomorphism" as in a system to be described in the next chapter? The earliest and perhaps clearest answer, at least as to the phrenologists' choices, comes from the 1825 work by George Combe. This well-argued text carries the name *The Constitution of Man*. The book went through many editions, and is superior, in terms of logic, clarity, and sensitivity to data, to anything written in later phrenology. Combe, as an original thinker, suffered the fate so common to scholarly folk: they need to be spared from the uncritical intensity of belief among their followers.

Combe, like many we have encountered, closely follows the lead provided by Immanuel Kant in his *Critique of Pure Reason* which appeared twenty-five years earlier. Like Kant, Combe rarely, if ever, shows an interest in individuals. Both men searched for the functional aspects of the universal mind that strain and categorize the information it receives. They are interested in the general, in the anthropological and psychological, in identifying faculties that are characteristic of human beings. Both argue that some faculties are 'given', while others are acquired through experience.

Combe's interest in phrenology began as a wish to separate 'moral law' from 'natural law' — God-given law from the organization of nature. He understood animal and human life to be alike in the organization of their minds, if different in the quality of their abilities. He accepted the view that mental faculties have evolved, and that faculties would be found to have evolved throughout animal life. He thereby accepted the idea of a 'mental ladder' of living forms, but included all lifeon the ladder. The view that animals, too, have a mental and ideational life is an unusual one for his time and for ours. Combe's interest in the mental faculty of subhuman animal life surpasses the interest of either Darwin or Wallace. Combe's phrenology, established over the sixty years in which successive editions of his book were in print, bridges the ideas of the Kant and the impending commercialization of the phrenology, commercialization which overshadowed the core ideas and led to the derision of phrenology.

Combe argues that the human mind has four chief ways of knowing: by feelings, by sentiments, by intellectual faculties, and by reflective faculties. The first two, he notes, are also used by animals. The intellectual faculties include the senses (e.g., hearing, taste) and through them we may know aspects of existing objects (e.g., form, size, weight). Intellectual faculties include notions of locality, number, and time. The 'reflective faculties' include 'comparison' and a sense of 'causality'.

Combe was an alert and opportunistic experimentalist, this method being a way of science that was to disappear from phrenology once its popularity (and the accompanying fees) was revealed. In 1835, for example, Combe visited the Newcastle Lunatic Asylum and examined the skulls of several patients without awareness of the nature of the notes on the personalities and temperaments of these people made by the staff. (Unlike Gall, who doing like work already knew the staff's viewpoints.) In this 'blind' experiment, Combe was interested in discovering whether his diagnoses based on phrenology matched those made previously by the staff. For example, for patient J. N., Combe found the animal organs to be large and found 'cautiousness' and 'destructiveness' to predominate. The staff noted, in comparison, 'animal-like', 'bad character' and suspected suicidal tendencies, finding 'hypochondriasis' and 'hope to be small'. Like comparisons were made with other patients and, later in the month, with inmates in Newcastle Jail, again with what Combe and the staff thought to be generally positive results. Combe's analysis, made with the availability of only a few minutes in which to assess the skull, compared favorably with notes made by the staff based on their long-term knowledge of the person.

Combe's attempt at experimentation was unusual; most explanations of phrenology before his time were pronouncements based on authority alone. Combe's work lacks, to our modern ways of thinking, precision of definition, experimentation with controlled conditions, and an adequate sampling of subjects. Exactly how well his predictions matched those of the asylum staff is unclear, and one wonders if the both staff and researcher were not ready to please by finding similarities. Regardless, it can clearly be recognized that experimentation, and hence, empirical evidence were held in regard, and that the standards of the time were met, if not exceeded.

General Body Types

The broadest idea put forth by phrenology was akin to the four types of temperament postulated by the ancient world: nervous, bilious, sanguine, or lymphatic. The ancient Greeks related the types to the humors of the internal organs, this idea an earlier representation of body-typing. Indeed, inspection of the color of fecal material remained a means of assessing temperament into recent times. Its use as a means of establishing the source of illness remains. It is a characteristic of a folk psychology that new representations come about redressed and fashionable; the older idea is then dismissed as unscientific, or whatever approbation works at the time, the newer is perceived as 'relevant', 'scientific', or 'up-to date'. So the idea of the humors came to be regarded as medieval (as if this were a pejorative), and the new phrenology qualified as scientific.

An 1889 publication from Fowler and Wells of well illustrates the initial body-typing procedure used in phrenological analysis. It follows the Greek model of identifying human beings by their humors, but adds some information about the body type that accompanies each temperament. They write:

"The Sanguine temperament, as its name implies, is dependent upon the constitutional predominance of the apparatus employed in the circulation of the blood to the heart, lungs, veins, and arteries, and is indicated by a form of moderate fullness, light or brown hair, blue eyes, a fair ruddy complexion, with a fondness for exercise, and a general disposition to active pursuits.

The Bilious temperament is marked by a dark-yellow or brown skin, dark or black eyes, a strong, bony frame, firm muscles, and rugged, prominent features. The Nervous temperament is marked by . . . a general delicacy and fineness of the body, thin hair, small muscles, a pale skin, a large head. . . .a decided tendency to study, to live in the realm of thought, to cultivate art, poetry, and sentiment, to dwell above the world of mere matter, and especially that which is gross and coarse.

The Lymphatic temperament depends on a predominance of the stomach, digestive apparatus and glandular system, and is manifested by a roundness of form, with soft and flabby tissues, and a slow, languid, circulation. The complexion is pale, the hair yellowish, fine and limp, and eyes watery. The brain is slow and feeble . . .and its mental expression, therefore, lacks spirit and vividness. " [5]

Four gentlemen of the 1880s display the differences in Figure 5.2. (The Nervous Temperament is the painter Mallais; the others are unidentified.) One charm of Fowler and Wells' publications is the use of line-drawings of well-known, then-living, people to illustrate their descriptions. No doubt this use added a sense of frisson for the reader.

Sanguine Temperament

Nervous Temperament

Lymphatic Temperament

Bilious Temperament

Figure 5.2: Four men, four temperaments; as based on their facial phenotype. The notion of four distinguishing temperaments is found in ancient Greek thought and is found today, for example, in Carl Jung's 'personality types'. That the structure of the body and phase might indicate temperament reappears in contemporary thought in 'constitutional psychology'. From Sizer and Drayton, 1889, p 19 and 20.

The second procedure of phrenology was a listing and measurement of the major faculties of the temperaments. These traits were judged by their presumed location within the brain by feeling the skull area corresponding to the area. The idea well-suited then-recent discoveries regarding the structure of the brain, just as it is today recognized that there are specific areas of the brain for short and long-term memory, for memory of melody, for recognition of pitch, for recognition of certain shapes, perhaps for memories of the voices of others, of faces of others (like the so-called 'grandmother cell'). The faculties of the mind suggested by phrenology included 'amativeness,' 'conjugal love', 'parental love', 'friendship', 'calculation', 'memory of place', 'planning', 'thinking,' 'sagacity', 'pleasantness', 'worship', 'prescience', love of applause', and on, and on.

The third method of phrenology was to predict the pattern of the behavior of individuals. For the practicing phrenologist, a trained look at the facial features, hair and eye color, and shape of the skull were as revealing as the multiple-choice test of personality given today.

To Know a Nose (and Eyes and Ears)

'Dr.' Samuel Wells (of Fowler and Wells, proprietors of the Phrenological Cabinet) writing in 1867 in what is the most detailed presentation of phrenology to its time, discusses the relationship between various body parts and temperament: the chin, the jaws, the checks, the mouth, and, most detailed, the nose. Having chosen the category 'nose', we might postulate a number of ways in which noses, as a class, could be measured, categorized, and distinguished. We might invent a triangular device that measures angles of noses, thereby enhancing the reliability of our work and making it appear more scientific; we might use brightness or color as a category; we might invent a hardness scale, just as we have invented a pH scale, a Geiger counter, and a Richter scale.

Given these many possibilities, Wells chose to categorize the noses of the world by ethnicity. One such category is the American ("mixed, sometimes Roman"): he writes, "All the combative faculties are well represented by their characteristic signs, but that of Relative Defense shows, in general, the largest development, and this agrees with our national character and national history."[6] Wells further tells us that "The German noses are marked by Apprehension and Inquisitiveness, not fully developed...Secretiveness and Acquisitiveness largely indicated, along with some intellectual abilities." There are compliments for the Irish, with slights at the recent wave of immigrants. The Irish have "Beautiful noses of the Greek and Roman type, but the class largely represented in the US, alas, of a lower form." Native Americans, it is explained, have noses of the Roman type; energetic, war-like; some are Jewish, he tells us, as is the Negro nose. Page 219 of this treatise is reproduced in Figure 5.3, as the combination of illustration and text may be required for the reader to grasp the correlations proposed.

Douglas Keith Candland

NEGRO NOSES.

The Negro nose is the Jewish or Syrian nose flattened and shortened. We may call it the Snubo-Jewish. This abbreviation, of course, takes away much of the force of character and penetration that belong to the physiognomy of the true Jewish nose. Fig. 314 is an outline of the nose of a Negro chief, and shows indications of considerable force, but does not depart from the general form except in being less flattened.

Fig. 313.   Fig. 314.

THE MONGOLIAN NOSE.

We have already spoken of the Mongolians as a Snub-nosed race. In outline of profile we observe some diversity, but the prevailing form is the Celestial, as shown in fig. 315. There is in all cases both a flattening and an abbreviation in horizontal projection, in comparison with the nose of the Caucasian. See the Chinese, Japanese, Calmucks, Tartars, etc., for examples.

Fig. 315.

NOSES OF THE PACIFIC ISLANDERS

Of the natives of the Pacific Islands, those nearest the old continent of Asia, and therefore nearest the old blood, are of the lowest possible mental and physical organization, little elevated above the low class of animals—kangaroo and the ornithorynchus.

Fig. 316.

Fig. 317.—New Zealander.

Figure 5.3: Phrenology, the judging of talent and character by phenotype, became more specific, in time focusing on the nose. By the time of the Chicago Fair, described in Chapter 3, such characteristics were assumed to represent racial characteristics, in turn to be distinguishable by the process of human evolution. Indeed, the theme of the Fair was "Human Progress" which described both crafts and machines and the progress of human evolution. Types of noses from Wells, *New Physiognomy*, 1867, p 219.

142

In his studies of human behavior, Joseph Simms, M.D., also took into account the appearance of many facial and bodily parts. Figure 5.4 illustrates Simms's examples. Here we see (1) how ones eyes presage a proclivity for polygamy and (2) how different shapes of the ears presage musical ability, or its lack. Simms bolsters his claim by using the eyelids of celebrities, surely helping the people of his day get the point. The eyes used are those of 'polyerotically small' Mrs. Osoli, who seems to have but one lover, if that, and that of Brigham Young, whose trait for 'polyeroticity large' was well known. The eye of the turtle-dove is like that of Mrs. Osoli (although doves appear to reproduce far more successfully than she did). The hog, by implication, has traits recognizable by Simms as similar to those of Brigham Young.

The text complementing the picture tells us that "The amount of love for the opposite sex may be known by the fullness of the eyes, and its quality by the shape of the commisures, or opening between the lids of the eyes. When the opening is quite almond-shaped, promiscuous love prevails in that form; if the commissure has great vertical measurement, the love is connubial... "[6] Lower on the page, we see that the ear of Adeline Patti, an opera star of the day, is small compared to the unmusical ear, the owner of which remains blissfully unnamed and musically untalented.

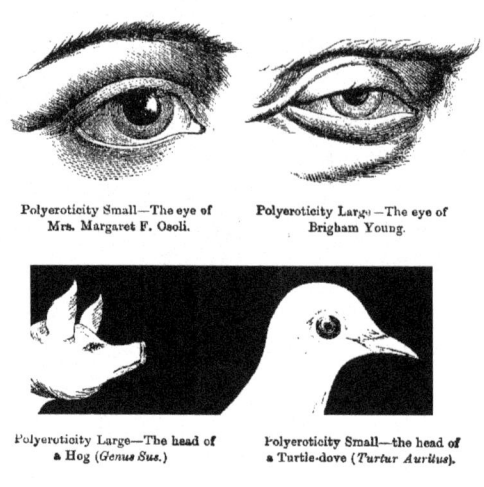

Polyeroticity Small—The eye of
Mrs. Margaret F. Osoli.

Polyeroticity Large—The eye of
Brigham Young.

Polyeroticity Large—The head of
a Hog (*Genus Sus.*)

Polyeroticity Small—the head of
a Turtle-dove (*Turtur Aurltus*).

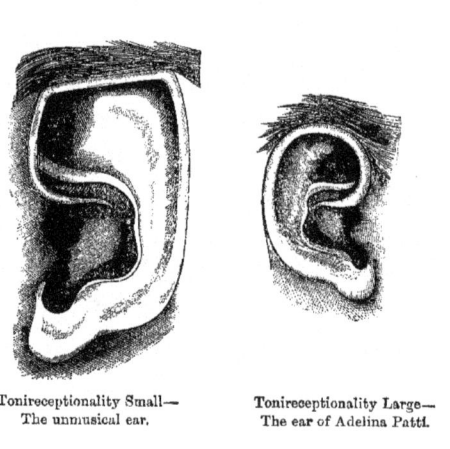

Tonireceptionality Small—
The unmusical ear.

Tonireceptionality Large—
The ear of Adelina Patti.

Figure 5.4: Derivatives from classic phrenology found various body parts to be predictors of talent, temperament, and behavior. This illustration, rather than focusing on racial differences, focused on individual differences of well known-people: opera stars, religious leaders, and leaders in society. Cesare Lomrose would use like comparisons to distinguish the criminal from the law-abiding. From Simms, 1887, p. 153 and p.163.

Simms was a pioneer in offering practical proposals regarding body-type and talent: he proposed that his 'system' of phrenology be used to select trades and professions appropriate for individuals. Simms had noticed that success is hereditary, so it followed logically that inherited features would predict success and failure. His analysis of racial and gender differences, in like fashion, argues for a 'progression' of both. On Simms's Mental Ladder, whites stand at the top of the five-tiered racial ladder, and men to the side and slightly above women. Such was a common British and American view of the late nineteenth century. It is not helpful to scoff at Simms's 600-page analysis [7] of the history of humankind. In particular, his view that behavior is mostly determined by hereditary genetics, but sometimes altered favorably or unfavorably by social custom and rearing, is popular today.

The Promise Unfulfilled

We return to Alfred Russel Wallace's work describing the successes of the then passing century (1899) and its promises for the future. He foresaw the power of phrenology to encourage the individual, the teacher, and society, to achieve his or her potential. He predicted that phrenology would be used to identify those in need of special tutelage or those who were mentally irresponsible. He saw the removal of that special brand of unhappiness caused by one's working at a trade or profession for which their talents or temperament were ill-equipped. Phrenology would lead to the selection of a vocation in which one could both excel and be content. Based largely on Combe's ideas, he claimed that phrenology:

"will prove itself to be the true science of the mind. Its practical uses in education, in self-discipline, in the reformatory treatment of criminals, and in the remedial treatment of the insane, will give it one of the highest places in the hierarchy of the sciences; and its persistent neglect and obloquy during the last sixty years [e. g., since 1840] will be referred to as an example of the almost incredible narrowness and prejudice which prevailed among men of science." [8]

He proposed that the idea of phrenology was supported by five principles, each of which has merit; namely, (1) that the brain is the organ of the mind, (2) that size is, other things being equal, a measure of power, (3) that the brain is a congeries of organs, each having its appropriate faculty, (4) that the front of the brain is the seat of our reflective faculties; the top, of higher sentiments; the back and sides, of our animal instincts, and (5) that the form of the skull corresponds so closely to that of the brain that it is possible to examine the proportionate development of various traits.

Wallace's commitment and prediction has, so far, not proven itself accurate: let us note, however, the data and premises on which the prediction is based. If they sound not quite right to the modern ear in terms of our current understanding of neurological matters, neither do they seem to be uninformed. The logic is compelling; the assumptions clear and testable by experiment. With changes in wording to bring a modern sound to the ideas, Wallace's analysis and argument might find approval by our contemporaries.

Publications on phrenology after Combe's work shifted to the analysis of individuals, a shift which in turn began the commercialization of phrenology and marked its -end as a respected mode of study. Alas, when combined with the commonsense view that success seems related to inborn traits, the discovery of the mental 'centers' was all that was needed to launch an industry. We should note that the industry was not much different from those that exist today in the purveyors of tests and measurements. True, the reliability and validity of many of these modern tests far exceeds those of phrenology. Perhaps it is also true that this is primarily because phrenology chose the wrong variables, as the shape of the skull does not now seem to be a good predictor of much of anything, but the statistical methods needed to determine this were not then available. What differs most, ironically, is that the phrenologists had a reasonable hypothesis about the workings of the brain to account for the traits they observed, whereas modern testers are often reluctant to propose any hypothesis regarding the source of traits such as, say, intelligence or neuroticism.

Toward the end of the nineteenth centuryas commercialization increased, books on phrenology became little more than pictures and descriptions of the personalities of well-known people, thereby being the "Lives of the Stars" magazines of the day. In other commercial directions, phrenology expanded, becoming 'holistic', in contemporary terms, to include beliefs about diet, general health, and education.

Fowler, for one, had a progressive outlook: Humankind's nature, he thought, was not fixed, but alterable by society and by an individual's taking stock of himself or herself and changing her or his habits. Fowler, writing in his tome of 1874, whose title by my count runs to 83 words — presaging what is to characterize the book itself — argues, among other matters, for a vegetarian diet for human beings. The reasons are familiar: human teeth are not carnivorously oriented (he writes), a diet of grain and fruit is more palatable and kinder to the environment, eating animals promotes animal propensities in people, and animal slaughterhouses offend, or should offend, our moral sentiments.

For the record, Fowler's suggestions for self-improvement also include careful and controlled levels of mastication with different chewing rates for different vegetables, the drinking of mineralized water and avoidance of 'soft' water; exercise, this daily (the 'Indian Dance' suggested is surely aerobic); use of alcohol only in wine, and only before bed; and the use of octagon houses for living space, the later an architectural style favored by Rudolph Steiner so as not to confuse the house spirits.

Fowler believed that women are far better architects than men, and he and his wife designed and constructed an octagon house in which they lived in New York State. Some such homes remain in the northeast and far western parts of the US. The purpose of the unusual octagon shape is to provide maximum light and air. The center is the stairway, this squared, allowing easy access to basement and all floors. The rooms are rectangular, with the ends created by the eight-sided shape used for closets and storage. The parlor, as it is rarely used, faces any direction except south, south being reserved for the living room and dining room, these being one large room, where the southern light is wanted in northern hemispheres. Never one to miss a commercial opportunity, Fowler offers designs and plans in his 1874 book.

A Fossil Record of Phrenology

Like any modern-day scientists eager to have his or her views heard, Fowler and others phrenologists appeared at a variety of functions where his ideas could be demonstrated and disseminated. We know little about how frequently and how convincingly phrenology was employed, how many people were doing it, or what training they may have received. It would, however, seem reasonable to assert that as large a proportion of the population accepted phrenology who now read their horoscope or social-advice columns in the daily newspapers. Estimates of the use of phrenology must be inexact, but without doubt it attracted a sufficient following that it would rival the popularity of self-help books.

Some saw phrenology, just as some see psychology today, as a progressive science, as a way in which talent could be identified and encouraged, thereby allowing individuals to reach the potential for which they were equipped. But in the hands of others, phrenology became anything but a progressive, or even a benign, force. It became a commercial empire. How many people made career decisions based on its predictions? How many found romance? How many changed lives from one path to another? In short, although the founders saw phrenology leading to human betterment, it is likely that its effects on individuals making choices in profession and love were for some, negative. gPhrenology, as a psychology, did not run the same course as Mesmerism and Freudianism: it led to no secret societies (that one knows of), no exile of the founders and promoters, no establishment of guilds and unions, no certificates of achievement. The notion that physical structure, especially that of the skull, was correlated with faculties of the mind was not disproved so much as it was replaced.

The idea of physiognomy predicting talent, character, and ability is so persistent that when one form of it comes to be seen as 'wrong,' the basic idea is reclothed and refashioned with a contemporarily acceptable design. So the promise of phrenology, that the organs (especially the brain) reflect degrees of talent, aptitude, and character, fails when we measure the physical aspects of the cranium — but, surely, the idea screams, physiognomy still is *right*: the skull may be the wrong place to feel, but we need merely to search elsewhere or, perhaps, peer more deeply to find the physical correlates of talent and aptitude. The most recent dressing has us peering into the brain itself, searching for the location of personality, memory, and the like. In between, study of the skull came to be replaced with a broader body-typing, and it is to that idea that our mental-fossil excavation now turns.

Chapter 6

The Mental Fossil of Physique and Madness

The idea that physical structure and mental structure are aligned and correlated has changed its representations of itself, its outer clothing, so to speak, over the years. Although it has been around as long as the written word, it did not merge with idea of 'human progress', nor gain a modern, scientific flair, until the advent of phrenology. Derided as that system may be, based largely on the limited and commercialized version promoted toward the end of its life, the underlying core did not disappear but re-appears in several ways to be described: in the study of temperament (Chapter 6), criminality (Chapter 7), and eugenics (Chapter 8).

Return to the Asylum

We begin by re-examining another of our mental-fossil finds. We return to Dr. Ernest Kretchmer (Chapter 2) who had been a student of Emil Kraepelin, [1856-1926] to whom we owe the invention of the nosology we still use, in large part, for 'mental disease'. [1] Kretchmer received further training in experimental psychology while a student of Wilhelm Wundt. Recall that in 1907, Kretschmer began his a program of measuring inmates at the asylum he directed. Kretschmer's work was concerned with the temperaments of those classified as 'mad', and it makes a serious attempt to relate body-type to temperament and the nature of the mental disease.

Douglas Keith Candland

He developed a number of 'constitutional schemes', each a set of physical descriptions, some subjective to be sure ('strong versus weak hair', 'strong versus weak genitals'). These were correlated with the kind of mental illness diagnosed, although Kretschmer's original plan was to relate the body-typology to temperament. In time, he became more interested in the relationship between body-type and the type of psychosis, for he came to believe that different psychotic conditions were correlated with different body-types. From approximately 1905 to 1915 Kretschmer and his asylum staff rated inmates' physiques by using a complex scale best described by reference to the "Constitutional Scheme" designed by Kretschmer. (Examples were shown in Figure 2.7.)

His interests became more refined. He investigated various forms of psychosis, such as paranoid schizophrenia, to determine whether body-typing could help reveal sub-types of the condition. Kretschmer shows an acute intuition and an ability to form generalities regarding the mind of the insane that hold up well over the century since his analyses began: his attempt to construct how the minds of the psychotic person are organized squares well with all we have learned since his times.

As the designers of the Chicago Fair believed that the ladder of civilization placed different races and cultures on different planes of civilized development, so did Kretschmer and his many followers. He shared this view with Galton and Darwin, but unlike Galton, he did not appear to believe that such differences were fixed genetically. The belief that human temperament is related to body-structure appears in the ancient European world in texts from Hippocrates' times forward. In more or less modern times, we found promoters of this relationship to include the phrenologists Franz Gall and J. G. Spurzheim.

By the beginning of the twentieth century, at the time of the Chicago Fair, anthropology was providing the possibility of evaluating the idea through its interest in the measurement of human variation. By the 1930's such measurement had become sophisticated. This was largely due to the influence of photography, which allowed such measures to be maintained and verified and the development of statistical tools, such as the Analysis of Variance, which permitted the statistician to 'partial out' effects and interactions previously undetectable. Even more importantly, inferential statistics allowed one to conclude something about a large population from a small sample of subjects.

A sample of Kretchschmer's published results is shown in Figure 6.1. The top table shows the number of persons, either manic-depressive ('circular') or schizophrenic ('schizophrene'), who were classified as of an aesthenic body type, an athletic type, and pyknic type, and so on. Without benefit of statistics, our eye tells us that of the 85 manic-depressives, 58 were classified as pyknic. In like fashion, we can see that the majority of schizophrenics (81) were classified as aesthenic. The clinical diagnosis and the determination of body types were made before the shown measurements were taken. The bottom part of the figure shows measurements of physical aspects of the body of both men and women of the aesthenic type.

Table 6.1: Body and State of Mind Types

	Circular	Schizophrene	
Aesthenic Type	4	81	
Athletic-type	3	31	
Asthenic-Athletic together		2	11
Pyknic-type	58	2	
Pyknic Misformed	14	3	
Diplastic	0	34	
Without Diagnosis of Build		4	13
total =	85	175	

Asthenic Types, Principal measurements [metric]

	Men	Women	
Height [cm]	158.4	153.8	
Weight [kilos]		50.5	32.8
Shoulder	35.5	44.4	
Chest	84.1	77.7	
Stomach	74.1	67.7	
Hips	84.7	82.2	
Forearm [circumference]	23.5	20.4	
Hand [circumference]	19.7	18.0	
Calf [circumference]	30.0	27.7	
Leg [length]	89.4	79.2	

Figure 6. 1. Measurements and prediction of psychoses from body-type. The diagnosis of 'Circular' refers to those who would today be diagnosed as manic-depressive, 'Schinzophrene' to schizophrenics. From Kretschmer, 1922, p. 28, from 1922, English edition, p. 21]

Kretschmer concluded from this data that

[1] manic depressives have an affinity with the pyknic body type,
[2] there is an affinity between schizophrenia and the body-types asthenic and athletic [the term 'athletic type' was later discarded],
[3] conversely, schizophrenia is rare among pyknics, and manic-depression unusual among aesthenics. [2]

Significantly, he noted that these "affinities" or correlations, as we would call them today, were stronger among men than among women. One reason may be that in his sample schizophrenia was far more common among men, while the frequency of the occurrence of manic-depressive psychosis was separated evenly between the sexes, as shown in Table 6.2.

	Circular	Schizophrene
Men	43	125
Women	42	50
Total	85	175

Table 6.2. Sex and Gender differences between manic depressives and schizophrenics. From Kretschmer originally, 1922, p.10 but taken from Sheldon's reproduction, 1963, p. 26.

Kretschmer's work became more widely known, and more widely used, in the 1920s and 1930s with the translation of his 1922 work from German into English. Alas, celebrity status would, in time, lead Kretschmer to make the same egregious move as did the phrenologists –

he began publishing analyses of the physical structure of celebrities, authors, poets, and artists, showing how their physical being was identifiable by the temperaments shown in their work or performances. These celebrities included von Humboldt, Locke, Mirabeau, Calvin and various racial types. [3]

The translation of Kretschmer's experimental work prompted like investigations elsewhere. Some such studies were done sloppily, as if the investigator knew the results that were wanted. Some were done with care; Some ignored standard experimental design, perhaps in the rush to make a point, while most of these studies were hampered, if not invalidated, by a lack of understanding of statistical methodology. Some of these, good and bad, we shall unearth shortly.

One other act of Kretschmer's should be noted before we leave our discussion of him and his measurements. During the Nazi years in Germany, when it was decreed that Jews could not hold professional positions, Kretschmer resigned as president of the Berlin Psychiatric Association in protest. His post was filled by Carl Jung. [4]

Our consideration of the next stage of the evolution of this mental relic returns us to the University of Chicago, whose first building was constructed to house the Congress of Ideas at the 1892 Fair. It was in Chicago that W. H. Sheldon [1898-1977] began the systematic measurement of body-types and their relation to temperament.

Photographing Physiques

Sheldon's researches began at the Universities of Chicago and Texas (at which he received both the PhD and MD) and continued at Wisconsin, Harvard, Oregon, and Berkeley. His goal was to relate physique to personality, temperament, and behavior. His practical reason was to assist the judicial and welfare systems in finding a prompt way to decide which young men would profit from parole and education and which from incarceration.

In Sheldon's initial study, for which the purpose was to standardize description of the human (male) form, 400 students were recruited from the University of Chicago, and, altogether, four thousand from unidentified midwestern and eastern universities. Writes Sheldon, "the racial element was disregarded, except that Negroes and Orientals were not included." His sample was of mixed Europeans descent, and was, he writes, approximately 10% Jewish, representing the racial characteristics of university students at that time. [5] So much for disregarding race.

It is odd and frustrating that so little useful information is to be found concerning the raters of the physiques, how they rated, and what the reliability may have been. Eventually, the rater is revealed to have been a long-time staff-member, whose measurements and ratings of the photographs "correlated so well with those of others" that it seemed wasteful to Sheldon to employ more folk at the task in order to assess reliability. It is announced, in one of the last publications, that she, the eventual lone rater, "reviewed and confirmed the somatotypes," but it would be helpful to know the training given the rater(s?) and, especially, the correlations between and among their ratings when done 'blind,' that is, without consultation with one another.

Sheldon and his colleagues, as noted in Chapter 2, developed a system by which the human physique was divided into one of three main types, these called endomorphy, mesomorphy, and ectomorphy. Each was further categorized based on the appearance of the upper, middle, and lower body parts using a seven-point scale (with four being neutral. The system yields 343 possibilities, such as 5-6-5, 2-7-3, and the like. A perfect mesomorph, for example, is a 4-4-4.

Further categories of interest included, for example, 'dysplasia', the 'g-index', and 'gnarled mesomorphy', 'viscerotonia', 'somatotonia', and 'cerebrotonia'. Dysplasia occurred when two of the three numbered characteristics deviate markedly from the third, indicating a poorly proportioned body. The g-factor measured conspicuous femininity in males, which was thought to indicate homosexuality, bisexuality, or hermaphroditism: it also distinguished college students from aviators. DAMPRATS are arty, theatrical, good dancers, and probably given to homosexuality. Gnarled mesomorphy is a quality "often found in peasant or 'low' stock of Southern European origin. It is most common among Italians and Jews. The bodies are strong, often extremely strong, but are usually heavy and cumbersome, with massive torsos and short legs. They are built close to the ground and suggest stunted, gnarled trees growing near the timberline. In its most extreme form, gnarled mesomorphy has a ghoulish aspect." (writes Sheldon). [6]

The invention of yet another identifier, the 't component' is revealing. When Sheldon describes its origins, he tells us something of his own background and personal qualifications, thereby perhaps something of the motivations that guided himself and his colleagues to undertake these measurements. Revealingly, Sheldon tells us:

"The t component. For those who have attended dog shows, horse shows, girl shows, poultry shows, or other competitive exhibits of livestock, the t component is an old familiar friend. It is merely the physical quality of the animal.

As a boy I was, quite naturally, trained to judge poultry and dogs, for my father was a competitive breeder of both. . . . By the age of 12 I had attended many different kinds of livestock exhibits, and was probably as competent a judger at most of them as I could ever expect to become." [7]

In a demonstrative study, the individuals photographed were members of a Delinquent Home (referred to as the 'Inn') and a chief purpose of the physical measurements taken was to assist in placement of incoming boys and 'deciding on a course of treatment'. The practical problem, as true now as then in like situations, was to:

"Acquire as rapidly as possible an understanding of the predicament of each new arrival and to respond as promptly as possible with help tailored to his unique needs . . . It was in our own defense that this method was introduced at the Inn. At the peak period, in the late 1930s, as many as 648 boys passed through the

Inn in one year . . . We had to determine [a] whether we could work constructively with the boy in our group setting, [b] what strengths, weaknesses, and potentials for growth he showed, and [c] whether the resources available either in the house or in the community could meet his needs . . . The somatotype, generated from a clinical photograph and supplemented by other anthropomorphic measurements, became a keystone in the edifice of diagnostic procedures and a basis on which to develop individual compensatory programs." [8]

The data were also used to predict the future kind and degree of criminality expected from a given youth. With such a prediction, it was thought, society could work to alter the outcome for individuals by education, nutrition, employment, a change in family, or whatever other means it might take to changes delinquents into productive workers. It should be noted that the term 'delinquent', as used when the study was performed, was a bit different from the use of the term today. These boys and young men were truant from school, regarded as difficult to deal with by their families, but few had committed the sort of crime that would lead to a incarceration sentence today.

When published, each picture of a delinquent boy was accompanied by many physical measures and a page or more of written analyses describing the subject's skills, abilities, personality, and temperaments. Like information is provided from photographs and ratings of military personnel (presumably the 'normal' sample). Combined, these yield a book that runs to 900 pages and contains an almost equal number of photographs. Reading through the book, it is difficult to see what the precise measurements have to do with the prose descriptions, for the predictions compare oddly and unexpectedly with the mountains of figures and graphs that accompany the physical measurements. It is as if the authors used the text as an opportunity to describe personality aspects thought inappropriate if expressed in the statistical material - as if too many data interfered with the descriptions Sheldon and colleagues wanted to provide.

To illustrate the system Sheldon and colleagues developed, I reproduce three photographs in, Figures 6.4, 6.5, and 6.6, followed by their accompanying description. I chose numbers 44, 125, 170 by opening the book to three different places at the beginning, middle, and end and found them to offer a range of the sorts of commentary presented. When the selections were made, I had yet to discover the truly important aspect of this work: Information was collected on these men, after ten years (1939) by Sheldon's colleagues and, after thirty years, by Hartl, Monnelly, and Elderkin, folk unrelated to the original investigation. Such follow-up studies are rare in social science and therefore this effort is all the more laudatory and helpful. Learning of the course and outcomes of their lives was a surprising opportunity.

We begin by reading the original analyses provided by Sheldon and colleagues, these made in the late 1930s; then we shall learn the course of the boys' lives during the next four decades. We are told that the boys were all 'in trouble': runaways, and truants ready to be placed in an "Inn" for shelter, education, and training. Edited and abridged accounts of the written descriptions and evaluations of 44, 125, and 170 are here provided. (Figures 6.3, 6.4, 6.5.)

My purpose in abstracting these descriptions and commentaries is twofold: First,
I want the reader to grasp the predilections of the commentator, the non-too nuanced themes that the commentator finds to be appropriate, for it is suggested that the descriptions say more about the interests of the evaluator than about the boys. Second, as these comments and classifications were used to determine the boys' educational and occupational future, we may ask two further questions: were the evaluations reliable, in that independent observers would observe like characteristics and, second, are the observations valid, in that they do, indeed, predict the boys' futures?

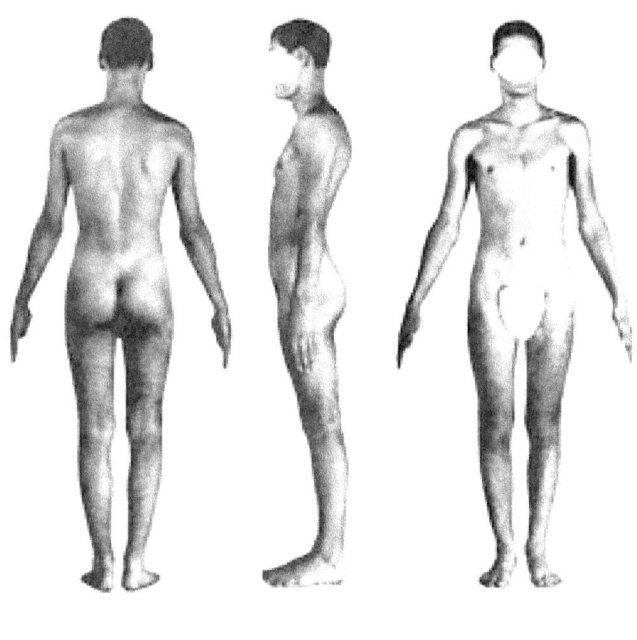

44

Figure 6.3: Number 44 is the first of three examples from 'constitutional psychology' Measurement of the physical characteristics was used, at first, to distinguish which delinquents could profit from institutional care. In time, the measurements were used to predict character and talent; later to judge the probability of future criminal behavior.

Thirty years later other investigators interviewed the surviving delinquents to determine what had become of their lives; that is, to measure the reliability and validity of the original estimates. The procedure (Chapter 7) was also used to determine criminality in general, used to measure college-aged men and women in the 1950s, and reinforces the mental fossil that body-shape determines social behavior. Delinquent youth number 44 was a truant, a petty thief, and an attempted stole car thief (also, reputedly, a 'sissy') wrote Sheldon, compiler of constitutional psychology. Photographed in the 1930's, described initially in Sheldon, 1949, p. 244 and 245,

Number 44. Sheldon writes:

"Somatotype 3-4-4. An 18-year old of midrange physique, two inches above average stature. Trunk ectomorophic. Heavy skeleton with comparatively inadequate musculature throughout. . . . High waste with feminoid narrowing of the middle trunk. Face highly feminine. Features too delicate, poorly molded, and acutely symmetrical. Hands and feet crude, weak. Coordination good in feminine sense. Graceful and willowy with girlish poise, he has a coy eye flutter and an almost lecherous baby-faced stare. Walks with the hip sway of a vivacious, tarty girl. Inept at athletics and combat.

Temperament: The behavioral pattern that of a highly energized but rather spoiled girl. He is a DAMP RAT with tantrums of perverseness the most conspicuous characteristic. He has too much energy, is restlessly somatorotic, yet his voice is well modulated and his articulation is perfect. He seems to love the voice for its own sake, loves to talk in the most cultivated manner of Tchaikovsky, Bach, Van Gogh—but never of Joe Louis, Truman or Li'l Abner. The content of his speech seems less important to him than the manner, articulation, and tonality of the utterance. Beneath the overwhelming gynanrophrenia there is almost paranoid suspiciousness, and somatotonia seems to simmer like a pot of beans on the back of the stove.

Origins and Family: Youngest of four, urban family. Father a French-Canadian who disappeared before this boy's birth and of whom little is known at local agencies or clinics. Mother Old American who is said to have had a delinquent history before marriage and shortly after this boy's birth was sentenced to one of the state reformatories.

Mental History: Achievement: Finished the eighth grade with no failures although considered a 'peculiar sissy.' IQ reports range from 85 to 103 . . .

Comment on Outlook: (Ten years later) . . . he has kept out of trouble, although he has not shown any signs of any major intellectual advance and he is of course the same DAMP RAT as ever. Two or three who know him state that he is decidedly not a homosexual, and on the strength of that evidence he is here classed as an example of the DAMP RAT syndrome who, in adult life, at least, has been free from homosexuality. Since one of the conspicuous secondary characteristics of the DAMP RAT syndrome is mendacity it is singularly difficult to collect accurate information on borderline homosexuality. Young men in college fraternities have reported encountering the same difficulty in collecting information on the sexual activities of their sorority classmates. I am told that there are some women, and perhaps also some homosexuals, who regard their sexual activities in such a light that in extreme instances they tend to mislead inquirers on the subject." [9]

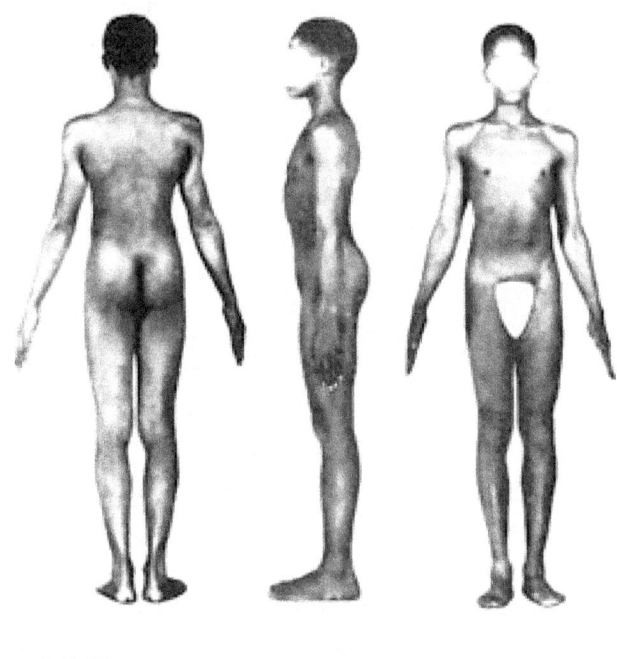

125

Figure 6.4: Delinquent youth number 125, a 'willful' and 'stubborn' truant. Photographed in 1930's, described initially in Sheldon, 1949, p. 489-491, re-examined thirty years later.

Number 125

Description: Somatotype 4-4-4. A 16-year old Negro youth of midrange somatotype and in inch above average stature. Immature physique. . . . Short trunk. A physique that will be heavy and soft in later life. Features feminoid and delicate for stock, finely molded and regular. Hands, feet, skin, and teeth all show excellent structure. From a Negro point of view, he has aristocratic stock. Eyes remarkably large and luminous. Coordination good in the graceful or gracile sense. Good dancer. Mincing, effeminate walk. Good at games in a minor way — plays softball well. Unable to fight.

Temperament: DAMP RAT pattern, or arty-perverse and theatrical. He has inexhaustible energy, with somatotonia predominant. Willfulness seems to provide him with his persona. Frequently noisy, with vocal somatorosis, he is often in trouble and when crossed is inclined to go over into a sort of hysterical rage which may be accompanied by a deafening screech. He will slap smaller boys but that is the extent of his combativeness. He is affected, arty, has a superior air. Frequently accused, probably falsely, of homosexuality.

Delinquency: Truancy and willful or stubborn behavior during the early years of school. No individual record of stealing, but at 15 and 16 involved with a gang which was responsible for extensive looting. History of epithetical and hairpulling warfare with parents between 13 and 16.

Origins and Family: Only child of a brief marriage, although both parents have numerous children by other marriages . . . He promptly rejected these parents [upon being reunited with them] and his real delinquency dates from the time when he joined them, against his will. "They are coarse, ordinary Negroes," he says.

Comment: Outlook probably good. He has the IQ, the health, and perhaps he now has the motivation for at least a high school education . . . He is a hysterical psychopath, although of mild degree, and superimposed upon the hysterical trend is enough gynandrophrenia to render him uneasy in almost any environment. Insufficiently DAMP RAT for homosexuality. I think, he falls somewhere between that fraternity and the House of Masculinity. With a good education and a mellow outlook, such a position is tenable enough . . . [10]

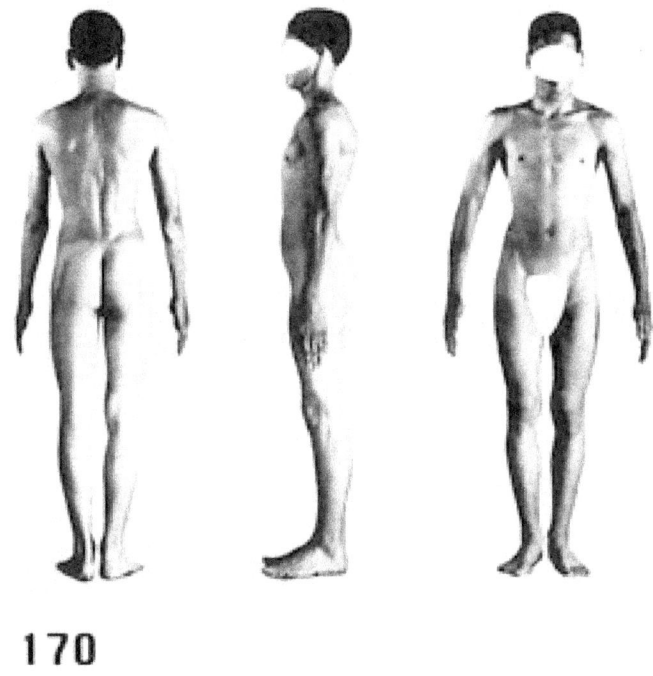

170

Figure 6.5: Delinquent youth number 170, a petty thief who dropped out of high school and 'refused to take anything seriously'. Photographed in 1930's, described initially in Sheldon, 1949, p. 626-628, refound and re-evaluated thirty years later.

Number 170

Description: Somatotype 2 1/2-4-4. A 16-year old ectomorph-mesomorph three inches above average stature. Wide shoulders. Bones comparatively heavier than muscles. High, peculiarly-slanted waist.

. . .The face has an overly keen, birdlike appearance. Hands and feet well formed. Coordination good, although somewhat feminoid. He moves quickly and gracefully, but with a feminine hip sway. Good at minor athletics and fairly good at baseball pitching. Not good at contact games like football and basketball. He is no fighter.

Temperament: His mind and most of his energy had got stuck, or focused, on the subject of sex. He had a 'friend obsession' said one psychiatrist. Friendly and sociable in a superficial sort of way, he seemed to have a new set of friends every few days. He touched all things lightly. Only when cornered and faced with the necessity of accepting some temporary responsibility did he show any signs of cerebrosis . . . There was no masculine pugnacity in him. . . he was airy, a trifle gynandrophrenic yet not quite arty. He bordered on DAMP RAT territory but was not quite a DAMP RAT and was certainly not homosexual. He had none of that peculiarly affected 'Oxford' speech which is often the password among DAMP RATS of the polite stratum 1-2-1.

Delinquency: Minor stealing between 12 and 15. . . . Quit high school at 15 and thereafter refused to take anything seriously.

Origins and Family: Extramarital, urban parents, father unknown. Mother a healthy Scotch-Irish woman of average physique, from a family of low economic status.

Mental history: Finished one year of high school with several failures which teachers called unnecessary. IQ reports fall between 95 and 106.

Comments: Outlook probably good in the sense that he is not a criminal or likely to require institutional care. Outlook poor in that he will fail to make the most of his natural endowment. He is likely to become an intermittently employed taxicab driver in South Boston. Probably he just barely escaped homosexuality. He has the sort of gynandrophrenia that the most persistent and incurable homosexuals show, but less of it." [11]

We do not know what decisions were made that altered these boys' lives, such as whether they were to be institutionalized or placed in foster care. We cannot turn the clock backward. We can, however, learn something about the accuracy of the predictions, for, as is to be reported, by other investigators sought and found these boys thirty years later. The interviews about how their lives developed provide, as we shall see, information on the accuracy of the predictions based on body-type.

Thirty Years Later

Social science is blessed when longitudinal information is available for comparison, for this opportunity provides a measure of the validity of the original predictions.. Such studies are problematic and difficult: living beings die, move, and become incapacitated, sometimes because of the institutionalization created by the original prediction. The original 'somatotyping' as the categorization made from the physical measurements is called, and the describing of the boys in the school was done in the early 1930s, a time of economic depression which, like times of war, put serious strains on families and social services. Such social services as were available were usually sponsored by churches and philanthropic groups: the idea that such service was a responsibility of government was not yet firmed. Thirty years later these men had lived through an economic depression, another world war, and through the tumultuous domestic readjustments which followed. The pictures used in that text were those from the original study;

The re-reviews were begun in 1958 and continued until 1960. A second review was carried out in 1968; a third in 1972. The follow-up procedure began by sending a questionnaire to the 200 men living in the Boston area. The reader is not told how many responded. The follow-up continued with in-person or telephone interviews with some of the men. The reader is not told the number or how they were chosen. Social service records were reviewed in the hope of finding information about the men (presumably laws regarding confidentiality would, today, prohibit some forms of this information -gathering). Using information from the questionnaires, documentation of the men's lives, and the interviews, participants were scored on a temperament scale. At the same time, the original 'somatotyping' categories were reviewed and, where necessary, revised (though whether this was done from the original photographs or based on the participant's current physique, we do not know).

The authors of the follow-up study found it worthwhile to add some new terms, and to redefine some that Sheldon had used. In doing so, they developed a systemic way in which to classify temperaments, its kinds and levels. They coined the terms 'endotonia', 'mesotonia', and 'exotonia' to describe three components of temperament (the similarity to endo-, meso-, and exomorphic body types should not be missed). "Endotonia," they write, "expresses the predominance of the digestive-assimilative function — the gut function. Briefly, 'endotonia' is manifested by relaxation, conviviality, and gluttony for food, company, affection, and social support."[12] Mesotonia, involving a predilection for "movement and predation.... is manifested by bodily assertiveness and a desire for muscular activity. When this second component predominates, the motive for life seems to be the vigorous utilization of energy. Mesotonics love action and power." [13] Finally, "Exotonia," we are told, "involves cerebrally mediated inhibition ... It leads to conscious attentionality and a substitution of symbolic ideation for immediate overt response to stimuli... This may lead to hesitation, disorientation, and confusion... [It is] manifested by (a) inhibition of both endotonic and mesotonic expression; (b) hyperattentionality or overconsciousness.... Ectotonics in a crowded society try to escape the painful consequences of the increased exteroception constantly surrounding them . . . There is an element of ecstasy in the heightened attentionality." [14]

Here, then, are the measurements and evaluations made thirty years later:

NO. 44 [1979] Second-Order Psychopathy, Uncomplicated

Description — Somatotype 4-4-4. Ht 71.5, Wt 173. A 56-year-old man of midrange physique. Trunk ectomorphic. Heavy skeleton with comparatively inadequate musculature throughout. A 4. G 41/2. High waist with feminoid narrowing of the middle trunk. SI1 and SI2 4. Features poorly molded and acutely asymmetrical. Hands and feet crude and weak. GStr and HStr 2, En 21/2. Coordination good in a feminine sense; inept at athletics and combat.

Temperament — [Previously] A DAMP RAT with tantrums of perverseness. He had too much energy and was restlessly memorotic. Voice was well modulated and he seemed to love the voice for its own sake. Talked in a 'cultivated manner' of arty things. Beneath the overwhelming gynephrenia there was almost paranoid suspiciousness. His arty pursuits were a sometimes thing.

[Now] As an adult he has had to plod for a living at a blue-collar level. He shows less gynephrenia, suspicion, and schizoid aloofness, but he is still cold and distant.

Delinquency — Childhood temper tantrums, truancy and petty stealing from foster homes; larceny and one attempted car theft. Committed three times to state correctional schools. No record of adult delinquency.

Family — Father a French-Canadian who disappeared before this boy was born. Mother of old American stock who was said to have been delinquent and was later sent to a state reformatory. [We now find that] She later remarried, had two more children, and was alive at last report. The boy was reared in foster homes under agency care.

As an adult he made his home with an unmarried uncle and aunt until age 44, when he changed jobs and moved to another community. Has never married or had children. Leads a rather restricted life. The uncle and the uncle's sister still live in the family home.

Mental history — Little formal schooling after grade 8. Low average IQ. Considered a 'sissy' by his peers. Verbal reactions were alert but superficial and revealed no evidence of mental integrity. No vocational plan. Some interest in music. His social presentation was that of a buoyant youth, coy and flirtatious. No further schooling or vocational training.

Medical — [Contains new information] Described as a large and soft child who grew rapidly and matured early. Never adapted well to his peers. Many psychiatric referrals with various diagnoses, such as emotionally immature and easily led psychopathic personality with paranoid trend. Had many infections, gland trouble, and was at one time suspected of having tuberculosis. No major medical problems as an adult. Teeth are soft and in poor condition.

Social history — [Contains new information] Frequent minor medical difficulties at the Inn with numerous complaints, which seemed to increase when school or jobs were suggested. There were accusations of homosexuality, through the boy himself never claimed or admitted such involvement. Showed improvement after leaving the Inn, got a job, and finally entered military service where he made a creditable record.

For the past 30 years he has been steadily employed in various capacities in factories. Has worked conscientiously and stayed out of trouble, with no car and no marriage. His principal hobby is an occasional game of golf. Also enjoys music. Now lives in his own apartment, attends church now and then, and has a limited social life.

Summary — Gynicasthenic mesomorph. The DAMP RAT syndrome has been subdued or overwhelmed by the necessity of earning a living at a blue-collar level. Good adjustment, and an acceptable citizen at his level.

Comment — Now near the middle of his sixth decade, he seems more subdued than when we knew him in his late teens; he keeps out of trouble, minds his own business, and lives as he seems to want to live -- alone. There was never any exploitation of his DAMP RAT pattern. Parasitism is far from this man's mind; he wants to earn is own way. Perhaps he knew instinctively that his low average IQ would prohibit success in that tumultuous competition. He is satisfied with his lot and does not look backward. There is no whining or complaining, no remorse, no blame. What second component psychopathy he has shows up in his aloofness, excessive privacy, and mild suspiciousness. [15]

No. 125 (1979) Alcoholic: Second-Order Psychopathy, Complicated
Description — Somatotype 3-5-31/2, Ht 70.1, Wt 160.

A 52-year-old black endomorphic mesomorph. Ectomorphic dysplasia in the distal segments of the arms and legs. Short trunk. Features and delicate for the stock. He has aristocratic blood which shows in the hands, feet, skin, and teeth. Coordination good; likes to dance, unable to fight.

Temperament — [Then he] Showed evidence of the DAMP RAT pattern: arty-perverse and theatrical. A willful fellow with much vocal mesorosis. When crossed, he went into hysterical rage. One clinic labeled him psychoneurotic with hysterical trends. Probably falsely accused of being a homosexual.

[Now] As an adult, Dionysian and irresponsible. It was rumored that he was addicted to drugs but this was never confirmed. Frequent binges of drinking. A random drifter for many years. The DAMP RAT element was submerged by the dreary derelict lift style. There was a later resurgence toward normality, and he seems finally to have balanced the conflicting elements in his temperament.

Delinquency — Truancy and stubborn behavior during his early school years. At 16 he was involved with a gang that participated in stealing and looting. He was the least guilty of the gang. As an adult six arrests for drunkenness, two for nonsupport, and two for breaking and entering at night. Sentenced to the house of correction on two occasions.

Family — An only child of black parents who were separated when the boy was born. He was reared by relatives for a time. The father remarried and the boy rejoined the family at age 12. He was unhappy in the home, and a running battle ensued between him and the parents. He later said that his troubles began at that point. The biological mother and stepmother are both dead, causes and ages at death unknown. The father is alive in his late 70s.

He was married at age 22 to a capable girl. They had two children, a boy and a girl, both of whom have had college educations and are now professional people. Sporadic employment history. He began to drink heavily, and after 10 years the marriage ended in divorce and the wife and children reestablished themselves elsewhere. He returned to an irresponsible vagrant existence, working a bit and cadging from his associates. Much drinking with the behavior of a derelict.

At age 48 he seemed to emerge from his alcoholic haze. Found work in a large hotel, quit drinking, married again, and has returned to a socially acceptable life style.

Mental history — Finished three years of high school. IQ average. Gifted in music and art. No vocational plan. Attended a machinists' school for three months while in service, but never made use of that training. Later took some courses in cooking under the GI Bill.

Medical — Negative findings at the Inn. He is said to have used addictive drugs while in service. Never verified. Had grand mal epilepsy while in service; hospitalized for a time, given a medical discharge and a partial pension. The last examination at a military hospital showed no signs of epilepsy and the pension was discontinued. Sporadically heavy drinking and moderately heavy smoking.

[Social history — He did well at the Inn but his artiness (DAMP RATism) offended some of the tougher boys. He eschewed the school program and found work in a defense plant for a time. Inducted into military service at age 17 where he did fairly well as a ship's cook. After World War II he re enlisted in the Navy at age 25 and served five more years as a cook on board ship. During this period he developed grand mal epilepsy, was given a medical discharge and a small pension, and tried several types of work. Meanwhile he married and had children. He failed to support, the wife and children left him, and he reverted to living like a vagrant. For several years he did menial tasks in low-grade bars and cafes. Between jobs he begged from his associates but did not ask for welfare. Paid little attention to his children. After about 15 years of this performance, he went to work, stopped drinking, and finally married a capable woman who had a job and some security. At our more recent report he was working steadily and seemed happily married.

A mesomorphic physique.

Comment: Good intelligence and fairly good health except for grand mal epilepsy, heavy drinking, and the use of drugs. A moderate degree of hysterical mesorosis. Some gynephrenic interference. A borderline DAMP RAT. This is another example of the internal warfare that can take place in a man between the male and female elements of personality. When we knew him at the Inn and up to the time of our first report, his outlook seemed good. He had relatively high-grade intelligence and an adequate structural endowment and seemed destined for a good outcome. But he began resorting increasingly to alcohol and became more overtly DAMP RAT. He seemed unable to live with the conflict within him. Discouragement with marriage followed and he jettisoned his wife and children and became increasingly alcoholic. As is typical of high gynemorphy, it took only a small quantity to intoxicate him. He seemed to be rapidly going down the drain. But at age 48 he turned around completely and stopped drinking. No question that his second wife was the key person in this reversal. What exactly there is in their relationship that led to this fortunate outcome we cannot say specifically. No doubt she is the stronger of the two and provides support, but he has some basic elements of quality. At the Inn we knew him as a capable person; and he has lived up to that early promise. His children reflect the element of quality. They pass the 'progeny test' with flying colors and seem to have surmounted their relatively fatherless early upbringing. [16]

Summary: His outlook was positive but, unlike the prediction, he never received further education and never amounted to much of anything except a failed marriage, of course. In hindsight, the staff find that he met their predictions. Further, the fact that his children (whom he failed to support) did well is seen to confirm that he could contribute to society.

No. 170 (1979) Second-Order Psychopathy, uncomplicated

Description — Ht 71.4, Wt 206. Died at age 39. An endomorph-mesomorph with bones heavier than the muscles. High, slender waist — gynic lower trunk. Features well developed but a little asymmetrical. Hands and feet well formed. Coordination good although somewhat feminoid; he was no fighter.

Temperament — At the Inn he was a minor mesorotic; he was maladaptively overactive, but usually inconsequentially. His mind and much of his energy was fixed on the subject of sex. Touched all things lightly. Only when faced with accepting responsibility would he become tense and apprehensive. There was no masculine pugnacity in him. A trifle gynephrenic but not arty. He was not quite a DAMP RAT and certainly not a homosexual. There was a leveling off of artiness and ectorosis as he matured. The mesorosis persisted but never to the point of civil delinquency.

Delinquency — Minor stealing from 12 to 15. Showed general irresponsibility. Quit school at 15 and thereafter refused to take anything seriously. The only adult offenses were two minor motor vehicle violations.

Family — Extramarital, father unknown. Mother Scotch-Irish, a healthy woman from a family of good standing. At last report, still alive in her 70s. Boy reared by relatives. At age 18 he married a competent woman who bore him two living girls and three living boys. She had several miscarriages. The children are all doing well. He owned a home.

Mental history — Finished one year of high school. IQ average. He appeared to be easily distracted, and the focus of attention fluctuated. No vocational plan. [Contains new stuff] He disliked mathematics. After the war he took flying lessons on the GI Bill and later apprenticed himself as a repairman of heavy industrial diesel equipment.

Medical — No significant pathology in adolescence. War nerves and shrapnel wound in military service. Called constitutional psychopathic state, emotional instability at one hospital; another called him combat fatigue; still a third said his condition was psychoneurosis, anxiety state. Routine physical exam showed moderate arterial hypertension. Given a medical discharge from the service and a partial pension. After service he gained weight, smoked heavily, and drank moderately. Then in his 30s he complained of spots before his eyes and dizziness and was found to have high blood pressure. At 39 he died of a massive heart attack.

Social history — He cavorted around the Inn and at summer camp like a setter puppy. Made several starts at school and at jobs but failed to follow through with any of them. In the middle of his seventeenth year he entered military service where he had one minor wound and several episodes of war nerves and combat fatigue. While in service he married.

After discharge from service at age 20 he was variously a truck driver, an airplane pilot, and a repairman of heavy equipment repairing of diesel engines and worked in this field until his death. Bought a home, raised a family, had a number of hobbies, but did not watch his health. He lived at a furious pace, and death came at age 39. He was called a loving father and a devoted husband.

Douglas Keith Candland

Summary — A tall endomorph-mesomorph with some gynic interference. Average mentality. Early history of irresponsibility. Later made a good marriage, learned a skill, and established himself as a first-rate citizen. Died early of a coronary.

Comment — The pattern was one of overload in both the somatic components and the andrio-gynic dimension. In his teens this high endowment was more than he could handle. He seemed to be in a perpetual turmoil and unable to decide where he was going. There was also a dependence need to be cared for. (sic) This may be why the military did not exert its usual stabilizing influence. It took marriage to a sturdy, competent wife to allow him to organize his life. He really was not a delinquent, but rather a confused boy with too much energy and sex drive. Plenty of hard work, which he thrived on, took care of the former, and his wife took care of the latter.

This man's adult behavior reflected his structural underpinnings. He was restless, jerky, over spoken — friendly, sociable, and superficial. There was a faint suggestion of the restive-affected-theatrical quality of the DAMP RAT — but only a suggestion. He might better be described as a mesorotic with a light touch. Unfortunately for him, he chain-smoked through every waking hour of the day. This and his burgeoned weight contributed to his early death from coronary heart disease. [17]

Summary. Recall the original diagnosis. 'Outlook poor in that he will fail to make the most of his natural endowment. He is likely to become an intermittently employed taxi driver in South Boston. This seems to have missed the mark. He became a soldier, and died of heart disease. He did not become a criminal or require institutional care.

The authors of the follow-up work also offer us a statement as to how to undertake the 'Psychobiography' described in their work. As this description is the most complete and, at present, the ultimate statement of the 'constitutional psychology', it is worthy of our consideration. Psychobiography begins with:

"[an] examination of the individual's heredity,' then moves to [b] the physical constitution and developmental history; [c] the medical record and what might be known of the physiological history; [d] the mental history including school achievement and IQ records; [e] the emotional, behavioral, or temperamental history; and the social history of family and community relations, opportunities, achievements, and failures. [f] In short, it should include an acquaintance with the persona through the vicissitudes of the individual's physical, emotional, mental, and social life. [18]

Reliability of Body-type Assessments

The notion that a science of living beings itself begins with a description of morphology begins with Aristotle: it is, of course, a defining characteristic of how we study living beings. That morphology is essentially genetic appears evident, although, of course, we know that nutrition and environment influence the body. Information about physical constitution and medical history may well have been intended to provide information as to how morphology has been altered by environment. The attempts to gather information about mental history appear to make the same assumption about mental traits.

The follow-up studies might be expected to tell us something of the reliability of the methodology. From the categories used, we now know something about the frequency of social success, psychopathology, alcoholism, gynephrenosis, mental 'insufficiency,' and criminality. From these data we can find answers, or partial answers—perhaps only suggestions — to some of our questions. For example, we learn that of the 200 men originally described, 46 were known to be dead at the time of the follow-up, while eight were missing and presumed to have died.

Three of these were suicides, according to the authors of the 1982 follow-up, two of whom were 'homosexuals who were not at peace with themselves or their sexual orientation,' [19] and one a heterosexual who castrated himself: all three were 'problem drinkers'. Such clinical data, however suggestive, do not tell us whether the frequencies differ from that to be found in a nondelinquent population. Indeed, they appear to reinforce the stereotyping of personalities, especially in regard to gender-preference, and to suggest questions about the relation of alcohol to human life.

The authors summarize their findings, in small part, thus:

[A] cluster of morphological components appeared in our other studies of criminal populations..... The chief component is mesomorphic body build.... all studies agree on the dominance of mesomorphy in the physique of delinquents.... [20] The development of alcoholism then will happen more readily in endotonic and mesotonic personalities. [Note the shift from discussion of body-type to 'personality'.] In endomorphs, in whom relaxed, indiscriminate amiability seems to be an overwhelming necessity for the personality, the alcohol habit and become a vice... [21] Although most homosexuals show signs of gynephrenic behavior, and some show signs of gynephrenic difficulties, the converse is not true. [22]

Despite the sophisticated statistical analysis, the same flaws in logic that accompanied the development of phrenology present themselves. First, the data are necessarily correlational; no causal relations can be, or should be, made. Second, there can be no control group, for the decision as to the persons' lives made on the basis their body-type necessarily precludes any other outcome. Third, and most telling, there is never an attempt to compare the data with the general population. It may be so that the majority of criminals are mesomorphic: It may also be the case that the majority of males ranging freely among the population are also mesomorphic.

The data that might interest us most are these: do any of the measures taken during the first study predict the men's behavior during the intervals? To our chagrin, the analysis provided in the follow-up study are chiefly based on data taken from the men at the time of the follow-up: correlations between these and the first study are not conspicuous. Sometimes the original prediction is carried out, especially in the achieved level of education, but as the data were used to determine the degree of education allowed, the prediction is self-fulfilling.

Often, the re-evaluation appears to select from thirty

years of living those aspects that loosely fit the prediction, setting aside the unpredicted successes. The evaluators appear to be most interested in homosexuality and alcoholism. When none is found, the assessors suggest that lying about such is common.

The variable that accounts for most of the variance found in the follow-up study is 'number of years of school'. As the authors note [23] we should not be surprised, as this variable undoubtedly contains, nested within it, additional variances created by family structure, income, and intelligence and because pathology itself is responsible for school failure, and, ultimately, for the length of schooling. The second and third variables, in terms of degree of variation accounted for, are medical insufficiency and endomorphy. Whether the degree of correlation is sufficient to establish the validity of the measure is unanswered.

If we are left unfulfilled in the quest to determine whether physique is correlated to temperament, despite these noble follow-up study, we should at least be happy for the example set out by our archeological find. These studies illustrate, at least for the purpose of our analysis, a very recent redressing of the physiognomic idea.

Social Science and Mental Fossils

The work from Kretshmer through Sheldon, roughly from 1900 to 1980, was done to show that physique can be judged reliably and that, once done, different physiques can be shown to reflect different temperaments and psychoses. The variables initially considered important regard race, gender, ethnicity, and religion. By the final follow-ups in the early 1980s, these characteristics are ignored: emphasis is now on intelligence, family responsibility, schooling, and psychiatric pathology. Different questions are asked, making it awkward to establish a statistical measure of the reliability and validity of the two studies, yet, the changing categories tell us that the structure of the idea had evolved.

What we experience from our review of these attempts is not much about body-type and behavior, but awe at the imperious ways in which an idea directs and controls the attention of experimenters and clinicians alike by controlling the nature of the questions they ask and therefore the ways in which they interpret the results found. If the terms of 'constitutional psychology' — DAMP RATS, for example, — are unfamiliar or amusing, why should contemporary terms be more accurate or less dependent upon their underlying assumptions?

One evaluation of Sheldon's life and work [24] refers to it as a 'noble experiment', an 'experiment that showed some, but small, validity.' Although the 'noble' aspect is arguable, the comment is a generally fair one. Our studies of ourselves are often incomplete, often flawed methodologically, and usually without the long-term assessment that the behavior of living beings requires for proper understanding. Can social science be predictive of behavior? And, if it can be, do we want it to be?

From the viewpoint of the putative ideas here being unearthed and examined, the question is not whether physique predicts delinquency, madness, or temperament, but why one should even ask the question. Why do our mental folk-psychologies accept the view that body-type predicts temperament? The notion that physique and temperament are related is among the oldest of fossils in our collection of human ideas of itself: what Kretschmer and especially Sheldon did to re-dress and re-clothe the idea was accomplished in part by taking advantage of new tools to measure variations statistically. But there is a tone to the work of both persons that cannot be caught by reviewing their findings or providing selected quotations. It is a tone that describes their sense of supporting 'Human Progress' by being able to predict character, temperament, and criminality, that same tone that in a more populist and progressive way pervaded the Chicago Fair.

Kretschmer, working in Germany in the early 1900s through the 1920s never suggests that body-type might be used to predict psychosis or its type: his technique wouldn't allow such a conclusion. He appears, instead and more basically, to be interested in why psychotics of one type should have a different body-type from psychotics of a different type. He suspects that a third, as yet unfound, cause is related to both physique and form of madness, and tentatively suggests the endocrine system as the place to look. He appears not to be interested in creating a 'better' society or in predicting behavior: he is interested in what causes the different forms of psychosis. Kretschmer understood the methodological workings of social science far better than Sheldon and his followers. Nevertheless, he clearly thinks that humanity will be advanced by the finding of such underlying causes, and that his work is contributing the progress, if perhaps only in the scientific and technological sense.

Sheldon's postulation of the DAMP RAT component is a bright warning signal. On the one hand, he knows aesthetic perfection in animals and humanfolk when he sees it (he tells us); on the other, he understands the cultural relativity of such a statement; and, on a third hand if we can imagine this, he backs away from the predictive use of his own findings because such prejudgment does not belong, he tells us, in a democratic society. While seeming to accept the view that body-type determines temperament and delinquency, Sheldon undertakes his measurements and analysis in the hope of providing more equal opportunity. This seeming slippage in logic is characteristic of our reactions to the idea of physiognomy: it is now nowhere more conspicuous that when connected to measurers of the brain and body.

Sheldon's tone becomes ever more pragmatic over the years of his research, and more sensitive to the fact that imperfect prediction can be used to support more than one goal. Perhaps because the cost of the projects required foundation and, later, taxpayers support through government grants, Sheldon and colleagues, like phrenologists, came to emphasize the practical gains to be made from the ability to predict behavior. No doubt, funding agencies are more interested in results that can predict criminal acts than in findings, such as Kretschmer's, that state frankly its inability to separate cause from correlation. Quick payoffs are supported; long-term searches for relevant variables, not supported. As the cost of research with human beings increases, perhaps the need to offer studies whose utility is evident to the current folk-psychology becomes overpowering.

That the idea of physique relating to character appears and re-appears differently dressed seems clear from our analysis of its recent development. Nonetheless, constitutional psychology at present seems to be nowhere

189

obvious or remarked on, even in the teaching of psychology. The near-century of Kretshmer-Sheldon studies, work that once rated many pages in psychology textbooks, is simply not commented upon today. No one has disproved the work, or even spent time debunking it. When such ideas seem to disappear, we need to be cautious. It is likely that the idea did not disappear at all, but has merely reformed yet again, morphed into some other form that disguises the basic core idea.

We now turn a bit back in time, to find an application of physiognomy previously mentioned, but then unexplored. We have taken the notions we found at the Fair and pushed forward. Now, we take a step back, and will be rewarded not only by uncovering intriguing efforts to look at the connections between body and mind, but also in finding a conflict between two of our big ideas, one that continues unabated to this day.

Can we predict who will become a criminal?

Chapter 7

The Mental Fossil of Predicting Criminality

Would it not be a great service to humanity if we were able to examine a pack of, say, ten-year old children, and, by some relatively simple observation of each, predict those who would become criminals? We might want to do so for our protection or we might think that by so doing we could invent strategies to help change the future of these folk. The benefit to society and to the chosen individuals would be enormous. The future criminal could be provided with whatever social cures might be available. Society would prosper by eliminating prisons. The judicial system and attorneys would want for work, or learn to be otherwise useful. Noncriminal individuals could feel secure, and persons so tainted with whatever genetic or cultural trait that leads to criminal behavior, could be provided by society with the means to overcome what would otherwise be an unhappy outcome for both the individual and society.

When, at the 1893 Chicago Fair, anthropology, neurology, and psychology first exhibited themselves as scientific fields of study and achievement, each emphasized how photography could provide a permanent, reliable, and unchallengeable, record. The earliest-made photographic measurements were those made of Native American bones, dinosaurs and animal fossils, such as those of prosimian primates, great beasts of the Rocky Mountain States, and the peoples of differing cultures and races living in the Americas. The excitement accompanying the collecting and measuring was motivated, seemingly, by a desire to explain humankind and animal kind's 'progression', an idea thought to follow logically from newly-found knowledge regarding evolution, one which complemented the core idea of 'Human Progress'.

The practical uses of photographic measurements of variation appeared soon enough, suiting well, if not also prompting, the progressive spirit that prevailed. This spirit also owed much to August Comte [1798-1857] and the French positivists, who among other thing, argued that statistical measurement could produce the kinds data needed to set the direction of human progress. Ways in which human progress could be enhanced were as obvious then as now: it would be a happier world if insanity, mental retardation, and genetic diseases could be eliminated. Some thought it would be a better world if propensities for criminal behavior could be identified, thus allowing early treatment of the putative criminal, rather like inoculations against small pox, to prevent the later appearance of unwanted behavior. Some people extended the list to include other traits deemed unworthy of advanced human beings, such as mental retardation, homosexuality, and certain skin pigmentations.

The Criminal Mind and the Insane Mind

Crime, by definition, represents behavior against the accepted milieu of a specific time and place, even though exceptions rule. Murder, in peacetime, is a crime; murder, during declared war, is not. What is the characteristic of the criminal that compels her or him to the actions that society finds wrong or repugnant? Is the criminal insane, as some societies believe? Is the criminal a person who chooses evil over good, as other societies believe? Is the 'tendency' to crime an inherited one? Does some aspect of the environment, such as unemployment, create or abet criminal behavior?

'Crime' and the 'criminal' are important ideas, much in need of a thorough archeolopsychological exploration, for all societies experience them, for all societies deplore them, and all search for causes: all wish to eliminate criminal behavior. We learned in the last chapter that criminality was the single most obvious category of behavior associated with body type. Correlational findings are, of course, no evidence for, and possibly very misleading evidence of, causal connections. Nonetheless, the search for relationships between body-type and criminal behavior was to be in the large-sample studies in the 1930s and later. Its sources appear to be from a triumvirate of sources: the belief in human guidance toward human progress, the philosophy of positivism, and the political progressivism we saw at the Fair.

The present-day understanding of 'deviant' behavior is two-faced. Both law and medicine attempt to define the difference without great success. We legally punish with imprisonment the behaviorally deviant for his or her acts, but the psychologically deviant with asylum. But much muck occurs when societies try to distinguish between the acts of the person who knows wrong from right and the person who does wrong while thinking such to be right. The woman who kills her husband for profit is regarded differently from the woman who murders her husband because the voice of God told her to do so, yet the husbands are equally dead. In recent years, the defense of insanity has been enlarged by psychological concepts: murder may be forgiven if the murderer was 'sexually abused' by the murdered; violence may be understood as a result of suddenly enlivened and recalled 'repressed memories.' Lack of memory of committing a criminal act may be due to 'repression,' rather than 'forgetting' or 'willful distortion'. The difference between mental illness and criminal behavior is blurred, in part, because law and medicine hear and see different expressions of the acts.

In 1958 George Vold [1] analyzed the problem of separating the 'insane' from the 'criminal' in terms of a governing idea. In his book on 'criminology,' he points out that both insane and criminal behavior are longstanding aspects of human social behavior, that both are alike in that they produce dangerous, disapproved of, and often outrageous behavior. [2] The same, of course, might be said for children, the mentally retarded, the medicated, and many nonhuman animals – anyone, that is, who the folk-psychology of the time considers to be not responsible for their acts.

Our present-day–western psychology regarding insanity is grounded in the McNaughten rule of British law, this of 1843, which states that the accused must be able to understand both the accusation and how to defend him or herself; if there be charges and a trial before either judge or jury, the person must be able to understand the nature of the accusation, and whether the behavior was 'right' or 'wrong'. Most important for our study of the mental fossil record of the mind is whether the defendant was guided by delusion; that is, whether the person is 'out of his or her mind', a situation not unlike the result of hypnosis or animal magnetism, previously described. Inherent in the consideration is the out-of-mind experience: it matters not what is real, but what the defendant believes to be real. The person who hears a god speaking to her or him is 'guilty' only of mental illness if this god is real to that person, yet 'guilty' of whatever crime is involved if the person could distinguish the 'rightness' or 'wrongness' of the act the god demands. Insanity becomes both an explanation and a defense.

Given their differing understandings of the mind and its workings, we cannot expect the legal system and the medical/psychological system to come to agreement for the benefit of either the society or the individual. The differences lead to profound confusions in the public consciousness. May the legal system inhibit delusional and hallucinatory speech? May the legal system require the taking of medication by a person unwilling to do so? May the legal system require surgery, therapy, or the like as an alternative to incarcerations? [3]

The Born and the Raised Criminal

Another set of contrasting ideas that influence our understanding of criminality, and therefore how we chose to deal with it, are founding the oft-maligned nature vs. nurture dichotomy. I will not bore the reader with a lengthy explanation thereof, for it is not one of the main ideas for which we are now digging. Nor will I, beyond this sentence, moralize on the obvious falseness of the dichotomy, for all things develop, and none are born purely of one influence or the other. Keeping the distinction in mind, however, will help us sort through the mental fossils we will uncover in this chapter, as their creators were themselves motivated by one side or the other of the nature/nurture distinction.

Each side of this dichotomy leads directly to one or the other of two proposed cures. One side argues that if we can find the trait that predicts criminal behavior, then the predictor is biologically set, thereby genetic, it is in the genes. If, on the other hand, the trait is acquired from the way in which society is set up, changing the rules and practices of society can alter or eliminate criminal behavior.

The most conspicuous mental-fossil example I have found emphasizing that genetics cause criminal behavior is from nineteenth century Italy, and features Cesare Lombroso's [1835-1909] work on predicting the criminal from physique, an idea we visited in the last chapters. His ideas appeared to be a practical outgrowth of phrenology and constitutional psychology while also appearing to lead to the progressive movement of using tests and measurements of individual variability to predict ability, talent, and future behavior.

On the environmental side, Lightner Witmer [1867-1956], who founded the Psychological Clinic in Philadelphia with the intention of using tests and measurements on children of differing backgrounds, had in mind the promise of such measures eliminating the effects of economic class, degree of education, and quality of rearing, so that "the slum child could have the same opportunity as the child of the rich". [4] A second period, this in the history of the United States, when attempts to relate criminality to body-type was influential, extended from the 1920s to 1950s. Because these studies were undertaken at a time before statistical techniques and computers were able to handle large amounts of data, they overwhelm the reader with literally thousands of pages of data correlating social and hereditary factors with criminality.

Diagnosing the Born Criminal

Lombroso's original theory, was derived from his examinations of the brains of dead criminals wherein he found a 'distinct depression' that he named the 'median occipital fossa' [5] After finding that nationality was a successful predictor of criminality, Lombroso refined his work by examining promising predictors, these presumably based mostly on the appearance of Italians. The experimental method he uses is to examine the frequencies of certain physical aspects among persons already identified as criminals: the failure to compare these frequencies to a control of noncriminals is, today, self-evident. Over the next fifty years of work, as his methods became more sophisticated, his understanding of the need for better methods became more acute.

Lombroso initially suspected that head size was a relevant predictor of criminal behavior, but found that this is not so: rather, head size is correlated to race and nationality (which, one presumes, are themselves intercorrelated). If we take the head shape of the race/nationality as average; however, Lombroso found the skull of the criminal is more spherical. ('Exaggerations of ethnic types. . . toward the spherical is related to criminality.') Within the evolution of Lombroso's work we see the reliance of physiognomy: Because he wants to show that one ethnic group is more prone to criminality than another, when head size fails to relate to criminality he tries again and, and finds that spherical shape correlates with criminality. Lombroso's data revealed to him how the following attributes predict criminality:

The Face: "Disproportionate size, a phenomenon intimately connected with the greater development of the senses as compared with the nervous centers. [The comparison to subhuman animal life is readily evident in this finding: ideas do help us find what we are already looking for.] The forehead and jaw protrude [45.7%] in criminals. [We do not know how often this is the case in a noncriminal control group.] The lower teeth project above the jaw."

Eyes and Ears: "Asymmetry of these is the common characteristic [of criminals]. The eyes and ears are frequently situated at different levels and are of unequal size, the nose slants towards one side, etc. This symmetry . . .is connected with marked irregularities in the senses and functions. [The development of the senses were thought characteristic of animal life.] The eyes show a 'hard expression,' a 'shift glance,' oblique eyelids [A Mongolian characteristic]. Handle-shaped ears (like chimpanzees). A protuberance of the posterior margin [Darwin's tubercle], a relic of the pointed ear characteristic of apes."

Nose: "Twisted, up-turned, a Negroid character in thieves, in murders, on the contrary, it is often aquiline like the beak of a bird of prey."

Mouth: "Lemurine apophysis [the comparison is again to monkeys, actually prosimians, in this example], a depression in the jaw where the canine muscle is to be found. This muscle is commonly found in the dog, [where it] serves when contracted to draw back the lip leaving the canines exposed."

Cheeks: "Folds in the flesh of the cheek which recall the pouches of certain species of mammals."

Teeth: "The middle incisors are absent, a peculiarity which recalls the incisors of rodents"

Hair: "The hair of criminals tends to mimic the hair color and pattern of the opposite sex. Dark hair prevails especially in murderers and curly and woolly hair in swindlers. Gray hair and baldness are rare among criminals. The eyebrows are bushy and tend to meet across the nose." [6]

Given this information, we should have little trouble identifying the criminal from the noncriminal, and perhaps in stipulating the nature of the criminal behavior. Figure 7.1 offers some photos provided by Cesare Lombroso as examples: perhaps the reader can supply the appropriate diagnosis; perhaps not.

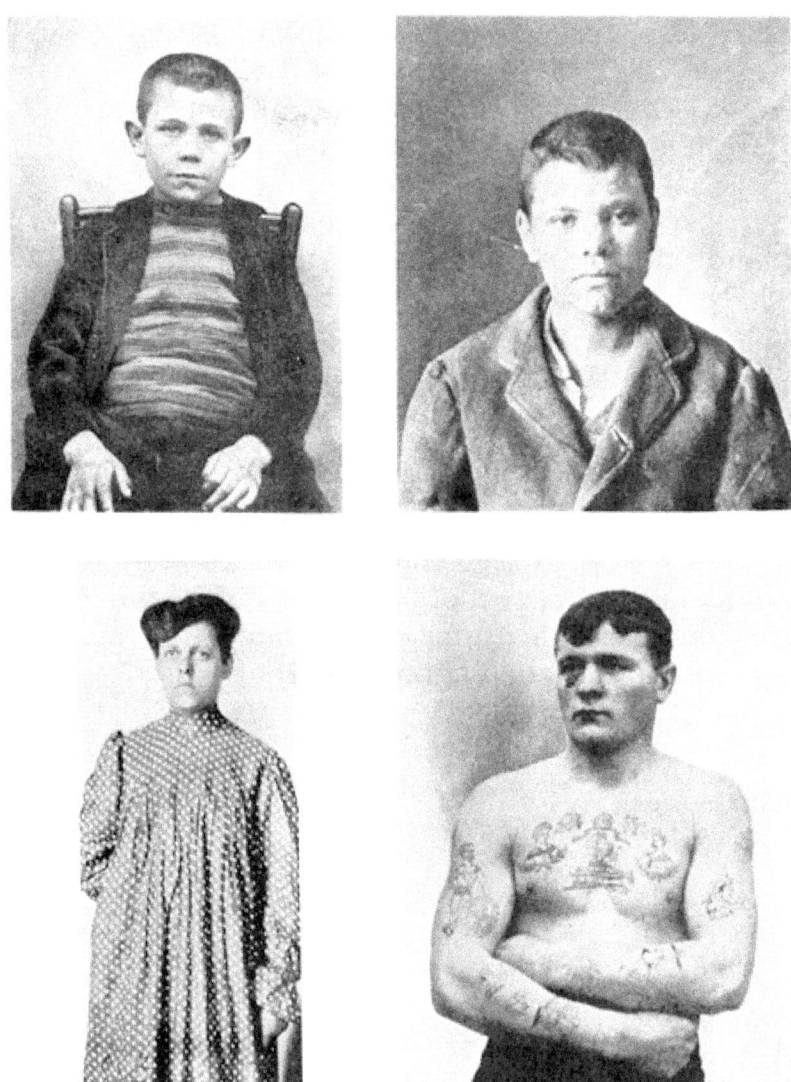

Figure 7.1: Examples shown by Lombroso of body-types and facial-types used to predict criminal behavior. Compare with Kretschmer's measurements and those of 'constitutional psychology' as promoted by Sheldon. From Lombroso, 1911, p 56, 114, and 120.

Lombroso held to a theory that criminal types were, by definition, a lower species, or subspecies of humanity. He believed that people have animal genes (as, of course, they do) and that the influence of these genes were verified when a person's appearance is that of certain animals. Some are rodent-like in features and carriage; some ape-like. Rodent-like folk are stealthy; ape-like, tricky and deceptive. The view that human beings could be separated into subspecies, racial-types, or animal-types, has not been an uncommon one in our lifetimes. Lombroso merely suggested that the lower types were more given to animal acts, some of which happened to be criminal. The criminal too is part of the ladder of human evolution, but rudimentarily so. The senses of the criminal, writes Lombroso, are more highly developed than is his intelligence; like animals avoiding or capturing prey, the criminal is designed for tasks that are habitual and instinctual; the civilized person designed for the finer thoughts that reason, thinking, and refined feelings allow.

Lombroso would eventually conclude that of the three types of criminals, the 'born criminal' type represented one-third of all incarcerated criminals. Treatment is lost on them, as they are genetically-controlled and genetically-fixed to commit their crimes. The born criminals are recognizable because they are recidivistic, begin their crimes in youth, and appear and re-appear before the courts. The slippage between 'born' and 'hereditary' remains unaddressed, no doubt because the words were interchangeable to many in those, and perhaps our, times. The born criminal is of major interest because 'he' commits not only most of the crimes, but because the nature of these crimes is aggressive, hurtful, and painful to the victim. Such a man was a creature of genetic defect, a throwback, an atavistic being. What, then, might society do to eliminate criminal behavior? As the environment is not responsible, changing the social environment is useless: culture is lost on the criminal. Biological problems require biological solutions, and, as we will read in the next chapter, such were tried.

Were we to interview folk today, obtaining from them honest answers about how they would describe the physical characteristics of the criminal, might we not hear some of the descriptions offered by Lombroso? Would we not learn of the predictive value of tattoos and the spacing of the eyes? Might we find, as well, the suspicion that certain body-types and facial characteristics, certain pigmentations, certain genders, are more prone to certain criminal acts? And might we find the view that any attempt to civilize the criminal is wasted time and money? The ideas seen here are certainly not extinct.

Charles Goring and his Well-Controlled Study

Charles Goring administered similar measurements of British convicts from 1901 to 1913 and compared these to those of students at Oxford, Cambridge, and to the Royal Engineers. Goring's work [7] was the true beginning of the use of experimental method in the social science of criminality. Goring didn't have the notion of a control group quite right in terms of his hypothesis: the members of the control group are not matched with those of the experimental group in any systematic way, but its presence is a major step nonetheless. His hypotheses were based on the work of Lombroso and, it would seem, on the earlier suggestions of phrenology. He expected differences to be found in head circumference, protrusions, and the like.

When convicts were compared to engineers, no differences were found between them. When criminals were compared to students, Goring found prisoners to be smaller in height and lower in weight. As Goring assumed such to be a matter of genes, his conclusions favored the notion of the inheritance of criminality. His twelve years of painstaking measurement and calculations, these in a pre-calculator and pre-computer day, seem somewhat marred by his insistence influences of nature, and his inability to see that height and weight, genetically constrained as they may be, are under the compelling and serious influence of other factors: nutrition, poverty, life-style, and education being among them. Nonetheless, let it be emphasized and credited, that when it came time to summarize over a decade of tedious measurements and calculations, Goring summarized the results with model restraint and care:

"Our results nowhere confirm the evidence [of a physical criminal type] . . . Both with regard to measurements and the presence of physical anomalies in criminals, our statistics present a startling conformity with similar statistics of the law-abiding class. Our inevitable conclusion must be that there is no such thing as a physical criminal type." [8]

Hooton and the Futility of Reasoning with Criminals

A serious limitation in the development of phrenology, as well as Lombroso and Goring's approaches, was the fact that data were based on data from only a few persons. Proper sampling methods were unknown, so data were limited to those who volunteered or were already distinguished in some way, such as being incarcerated. By the 1920s, two shifts resulted in the ability to measure large numbers of peoples' mental and physical attributes: the new ease of travel and the ability to contact large samples, whether individually or by photograph, and the newly worked-out statistical devices which permitted large-scale correlational studies.

An instructive such study was engineered by E. A. Hooton who received funds from a group of leading foundations, including the Social Science Research Council, the American Academy of Arts and Sciences, the Rockefeller Foundation, all respected major contributors to human betterment. Hooton and his many colleagues, some of whom will re-appear in our discussion of eugenics, had in mind gathering data in such a way and in such amounts that three distinct volumes of results could appear. The folk-psychology of the times clearly understood skin pigmentation to be a major determinant of criminal behavior, as these three volumes were so organized. The first volume reported on 4,212 White prisoners of native parentage in prisons and reformatories, and a comparable civilian sample. A second volume would be concerned, he promised, "with the material relating to native Whites of foreign parentage by nationalities, foreign Whites, and criminal and civil insane". "The final volume," he again pledged, "will be devoted to the study of County Jail prisoners, Negroes, Negroid, and Mexicans." [9]

Hooton was a man of whimsy and humor, as he also published little books of after-dinner speeches on apes and men [10] in which he opines how the races of humankind are contiguous with the great apes. However, in his practical work, his sense of whimsy sometimes detracted from the serious purpose to which the results were to be put. For example, he begins *The American*

Criminal by warning that, "It seems improbable that any reader, however hardy and enthusiastic, will be able to endure a complete page-by-page perusal of this book." [11] He is accurate both in his prediction and in his correct use of the word 'perusal'. The text itself is but 309 pages; the tables, taken together, double that and more. These are necessary, Hooton tells us, because, "If there is anything to the idea that criminality is predictable, all possible combinations must be correlated. Social science works by establishing, first, as many correlations as there are possible hypotheses, then examining those that appear to be most predictive." [12]

The many tables tell us, as example, of the percentage of these 'White prisoners of native parentage' who fit into different employment skills, indices of their breadth to height ratio, a facial index, a 'zygo-gonial' index, the reason for imprisonment, a nasal index, an ear index, hair quality, thickness of lips, wrinkleness of cheeks, gonial angles, and teeth caries, all by US state, both of birth and residence. We also see head length, sitting height and type of criminal offense, cephalo-facial index, tattooing, asymmetry, eye color, and body-build, all organized by the type of crime committed (or, more accurately, the crime they were accused of when incarceration). The methodology promised something heretofore impossible: large samples of people, information about many of their physical and mental characteristics, a freedom on the part of the investigators to learn details of family life and genetic background, and correlational analyses, the task made all the more awesome by the lack of computing machinery and, therefore, the necessity to employ statisticians working by hand.

Of the thousand of correlations and tests of significance engendered by this study, Hooton and colleagues chose to emphasize only certain ones, about twenty. These all demonstrated that prisoners have a lesser quality than the oddly-derived control sample. For example, "low foreheads, high pinched nasal roots, nasal bridges and tips... excess of nasal deflections, compressed faces and narrow jaws, fit well into the picture of general constitutional inferiority." [13] Note the slippery transition. The attempt to measure of body-type has now resulted in presumed measurement of 'hereditary inferiority'. Further, Hooton's emphasis is clearly not based on data alone – there are literally thousands of correlations reported that show greater effects than those chosen for discussion. Presumably, correlations that did not conform to the stereotype presented were understood as not relevant to the hypothesis.

Although the connections found have the appearance causal ones, we, as readers from the future, must remember that Hooton's method was strictly correlational. In contrast, Hooton and his colleagues often slipped by confusing correlations with causes. A misunderstanding of the meaning of individual statistically significant differences in studies using large scale correlations persists today, to the delight of the news when it reports that a study from this or that famous university shows this or that substance (carrots, wine) to predict (or diminish the probability of) whatever (cancer, obesity). When these correlations are reported as 'significant', what is meant is that the correlation has a probability of .05 of occurring due to chance — meaning that $1/20^{th}$ of all such correlations will be spurious ones. Given enough correlations — and Hooten reported literally thousands of correlations — one 1 of 20 will be found to be significant when they are not (and let us not forget that at least as many correlations may be found not significant, when real relation is present).

Hooton is not vague about the categories he thinks useful: "The object of the present investigation", he writes, "is to ascertain whether criminals differ physically from law-abiding citizens of the same race, nationality, and economic status, and if so, why." Having set nationality and race as the chief variables for investigation, not surprisingly, he finds these variables to be explanatory:

"The unpalatable fact of the matter is that it is a relatively futile procedure to tinker with the machinery of criminal justice, to strengthen and wash the hands of the police, to reason with the criminal or to pray over him, if the germ plasm which produces that criminal is scum." [14]

Many of the time recognized the flaws in Hooton's methods. Vold [15] listed the two major criticisms as:

(1) Prisoners are not a representative sample of criminals. They are, of course, merely a sample of persons arrested, charged, and convicted. The data are, thereby, a study of prisoners and convicts, not of criminals or criminality.

(2) The control group must be appropriate to the hypotheses being tested. Hooton's 'control' group was composed of folk whose work or habits made it possible to compel them to be measured. Those selected to represent the 'normals' included firemen, state militiamen (said to be uncooperative), mental patients and hospital out-patients (who may, or may

not, have been told why they were being measured), and Harvard students (then as now imprisoned and sentenced to do the bidding of their 'superiors'). Hooton further notes that selection avoids 'Negroes'. The most curious putative control groups were those trying to use a 'bathing house' at Revere Beach where potential subjects were approached while changing clothes; the recruiting technique was "not particularly successful, yielding only a few subjects." [16] It was unclear, even to those of the time, how this hodgepodge represented a proper contrast to imprisoned criminals.

We may add to the list of problems that measurers knew whether they were examining a prisoner or a normal 'control' and they also knew the hypotheses, hence what result was expected. There is no reason to suspect rascality, but we human beings do have a way of finding data to support our beliefs. Modern experiments should use a 'double-blind' technique, in which neither tester or testee know the group they represent, this to avoid the subtle but often impressive accumulation of data that inch by inch comes to serve the anticipated result. [17]

Another Try: Sheldon and Eleanor Glueck

Sheldon and Eleanor Glueck, both attached to the Harvard Law School, undertook the most reasoned, thorough, and artful study on the factors that might predict delinquent and criminal behavior. In 1929 they published an article showing the frequencies of certain aspects of the background, physique, and intelligence of a set of delinquents and presumed control persons. The work was first followed-up in their 1940 book *Juvenile*

Delinquents Grown Up. Further data and analyses followed in 1950 with *Unraveling Juvenile Delinquency*, and in 1959 with *Predicting Delinquency and Crime.* The later was a prefaced by Earl Warren, then Chief Justice of the United States, whose influence as a leading California attorney and long-termed governor will appear in the chapter to follow on eugenics.

The Gluecks work, which is dedicated to Dr. Hooton, generally follows his pattern and methods. The major difference in theoretical style is that the Gluecks' hypotheses led them to use correlations as predictors, while Hooton used them as explanations. The major methodological difference is that the Gluecks had a better understanding of the use of control subjects; they matched delinquents and controls in terms of 'ethnicity'. This primarily controlled for ethnic origin, but, importantly and not incidentally, also often controlled for religion, family structure, value placed on education, and income. Although the control for 'ethnicity' adds great validity, we should not forget that those aspects that correlate with ethnicity may thereby become overlooked, undervalued, and uninvestigated.

The first report of the Gluecks work emphasized the presentation of 'prediction' tables, these intended to allow judges and counselors to predict the likelihood that a particular delinquent could be retrained to avoid crime. They believed that a society armed with these statistical predictions could determine the appropriate disposition of delinquent persons: Prison? Reformatory? A foster family? To determine the factors that contributed to criminality, the Gluecks' established correlations between recidivism and the factors they thought to be relevant.

Three kinds of predictors were suggested by the Glucks: 'social prediction' (such as, whether native born or degree of education attained), descriptions of personality from the Rorschach test (this a 'projective test' of one's fantasy and imaginative abilities, thought by some to describe and predict traits of personality), and third, predictions made through a psychiatric interview. An example is shown in Figure 7.2. Here the three kinds of prediction are shown in the columns, each separated into a characteristic of a delinquent contrasted with a characteristic of a nondelinquent. Further, delinquents were categorized as typical or atypical, depending on the nature of their anti-social behavior. An social delinquent, for example, scores higher ("H") on 'feelings of not being taken care, higher on 'common sense' (as shown by the Rorschach), and higher on strength of handgrip and 'originality' If we recall the data collected by the 'constitutional measurements by Kretschmer and Sheldon, and compare their techniques to those used by the Gluecks, we see the improvement in behavioral research made possible by the newer understandings in experimental design and statistics, not a few of which were invented by eugenicists for just such purposes.

Table 7.2

Three types of measurement (social/economic background, application of a Rorschach test, a projective test designed to measure unconscious thoughts, and a psychiatric interview) are used to predict the qualities listed as 1-23. 10:11 AMhe participants are separated into those who are delinquents (Del.) and those not (nondel). Low indicates, as for example in Factor 1, a low IQ; H a high measurement of the quality. From Glueck and Glueck, 1959, p 166-167

Factor or Traits	SOCIAL		RORSCHACH		PSYCHIATRIC	
	Del.	Nondel	Del.	Nondel	Del.	Nondel
1. High IQ		L		L		
2. Poor Health in Infancy		H				
3. Restlessness in Early Childhood:						
4. Enuresis					L	
5. Handgrip				L	H	
6. Genital Under-development						
7. Originality					H	
8. Banality						
9. Common Sense				H	H	
10. Intuition						
11. Phantasy						
12. Social Assertiveness		L			H	L
13. Defiance				L	H	
14. Submissiveness				H	H	
15. Ambivalence to Authority		L		H		
16. Insecurity				H		
17. Feeling of Not Being Taken Care of				H		
18. Fear of Failure				H		
19. Resentment				L	H	
20. Hostility/Suspiciousness				L	H	
21. Destructiveness				L	H	
22. Defensive Attitude				L	H	
23. Marked dependence				H		

The Glueck's's views are similar to Hooton's in the variables they suspect to be predictive of delinquent behavior. Genital development, defiance, submissiveness, hostility — all appear on both lists of factors. We may learn something about the probability of delinquency from the data, but more importantly for us as archeopsychologists, we learn about how the beliefs of the time determine both the factors chosen for study and the interpretation of results gained thereby.

Consider the 'predictive score' for twelve-year old Frank, whose background variables are recaptured in Figure 7.3.

Table 7.3

The case of Frank. The goal is to ascertain the probability of recidivism. The factors are used to provide a statistical prediction for the use of judges, parole officers, and the like. . Modified from Glueck and Glueck, 1959, p. 102, 103

Predictive Factors	*Classification*
Birthplace of Father	Ireland
Birthplace of Mother	Ireland
Time parents were in the US	Six years
Religion of ParentsBoth	Catholic
Number of children in Family	Eight
Discipline by Father	Poor
Discipline by Mother	Firm but kindly
Conjugal Relations of Parents	Fair
Affection of Father for offender	Warm
Bad Habits in Childhood	Some
Age at Onset of Antisocial Behavior	Five years
Social Retardation	Two years
School Misconduct	Truancy and other
Time between Onset of Antisocial Behavior and First Arrest	Seven years

Prediction Scores for Each treatment or Post-Treatment Period:

Probability of Recidivism if

Given Straight Probation:

Birthplace of Father	53.7 %
Discipline by Father	45.5 %
Discipline by mother	15.4 %
School retardation: two years	62.7 %
School Misconduct: Truancy	62.1 %

Assigned to Correctional School:

Number of Children in Family	41.9 %	
Moral Standards at Home	42.3 %	
Conjugal Relations of Parents		48.1 %
Bad Habits	45.7 %	
Time between Onset of Antisocial Behavior and First Arrest	43.2 %	

Placed on Parole:

Birthplace of Father	89.7%
Birthplace of Mother	86.7%
Discipline by Father	63.9%
Discipline by Mother	50.0%
School Misconduct: Truancy	72.2%

Douglas Keith Candland

[For succinctness, three other possible courses of action are omitted.]

Summary of Frank's Prediction profile

Proposed Treatment	*Rate of Good Adjustment*
Straight Probation	74.0%
Probation with Suspended Sentence	37.8 %
Correctional School	34.5 %
Parole	15.4 %
Five Years after Completion of First Sentence	72.8 %
Completion of First Sentence	16.9 %

The predictions for Frank's future represent, of course, an important aim: everyone gains from accurate assessment, especially one that can distinguish the outcomes of various forms of treatments. From these correlations, the Gluecks are able to estimate that Frank's chances of good adjustment during the next fifteen years is 0.6 out of 1.0) ; that is, he has a better than even chance of staying 'out of trouble'. The odds suggest that the chance is better than even that he will do acceptable work in a correctional school and better than even that he will perform adequately if on parole, rather than being imprisoned. Although in later works, the Gluecks made it evident that they did not believe that these tables of results were suitable for decisions regarding parole, the fact is that the initial work described these data as predictive – one would be hard pressed to therefore not think it useful.

The Gluecks

The Gluecks, with the help of Sheldon, later classified young male delinquents according to their body-type following the procedures described in Chapter 6. After examining the body-types of these 1,000 folk, and calculating the percentages of this and that variable to be found among the several body-types, the Gluecks seem frustrated, but ready with fresh ideas. They write "The evidence of the present research, added to that of *Unraveling* [their second book], does not lead us to the pessimistic conclusion that persistent delinquency springs largely from the germ plasm and is inevitable." [18] In modern terms, they wish us to be happy they lack the ability to conclude that criminal behavior is genetic. Some correlations, however, were promising.

Having found that mesomorphs, like Irish-Catholics, are over-represented among the delinquent group, the authors offer an explanation that reveals a shift in the physiognomic idea. Rather than claiming that body-type is linked to crime through a predisposition for crime itself, they say:

". . . the basic reason for the excess of mesomorphs among delinquents as compared with other body types may well be that boys of predominantly mesomorphic physique are endowed with traits that equip them well for a delinquent role under the pressure of unfavorable sociocultural conditions; while endomorphs, being less energetic and less like to act out their 'drives,' have a lower delinquency potential than mesomorphs." [19]

Douglas Keith Candland

The shift from discussion of prediction to the more conservative discussion of potential is notable. So too is the suggestion that criminal behavior may be the result of aptitude put in a bad situation, rather than an innate tendency towards such behavior.

Spiritism and Criminality

Another view of criminality was taking hold at the same time social science started its large-scale investigation of the criminal. It was promoted by a group of people disinterested in statistical methodology and focused instead on the criminal personality. They thought that criminality was caused by unconscious conflicts or, perhaps, on the way in which the personality of individuals developed. Such categories as ethnicity or body-type were simply understood as the wrong place to be looking.

The Gluecks touched upon this when they compared the accuracy of 'statistical prediction' to 'clinical prediction'. The comparison is informative the alternatives represent two different fossilized understandings of ourselves. Do we trust the averages and deviations produced by measuring large numbers, or do we trust the detailed and refined work gained by in-depth analysis of a few cases, or even a single case? The Gluecks favored statistical prediction. Let us now examine the times innovations in clinical analysis of the criminal.

These can be divided into to three main parts. The first focused on the use of projective tests to identify the criminal. The second used large-scale statistics, but was focused the measuring the personality rather the body, and hoped from there to make the jump to prediction and treatment. The third looked to Freudian psychoanalytic theory to explain and heal the criminal psychy.

The Rorschach

A projective test is one in which the test-taker presumably 'projects' her or his imaginings onto stimuli that are intentionally vague, ambiguous, or meaningless. The Rorschach test is the ink-blot test so often seen on television and in movies. The results of such tests are analyzed by the administrator; they do not yield a ready numerical analysis, as do many other psychological tests. An advantage is that the subjects' answers are not forced into categories previously selected; thus the opportunity for variation is enhanced. On the other hand, the relative elasticity of the categories hands the interpreter the opportunity to use whatever categories she or he might think valuable. There remain serious questions about the reliability and validity of the Rorschach, as there are about all projective tests. Nonetheless, generations of clinicians have found it more than useful. In more recent times, any tests requiring lengthy one-on-one administration have lost ground to alternatives that can be administered more rapidly, often to groups at a time, and which produce numbers that can be compared more readily, if not more accurately.

Taking the results of the Glueck's study to be illustrative: Projective tests found that nondelinquents, as a group, showed a somewhat higher frequency of 'neuroticism,' 'psychotic trends,' and suggestively 'conspicuous pathology', while delinquents showed higher frequencies of 'asocial, primitive, poorly adjusted, and unstable traits,' 'psychopathology, and 'organic disturbances'. [20] Even so, these differences are in small percentages. It was also found that the non-delinquent controls exceeded the delinquents on 'feelings of insecurity and anxiety' and of 'not being wanted or loved.' These latter findings suggest a trait we would expect to be reasonably prominent among criminals; namely, a reluctance to examine one's actions and feelings. [21]

Personality Tests

Other psychological tests have been used to predict the probably-criminal. The The Minnesota Multiphasic Personality Inventory (the MMPI) is, perhaps, the most frequently used such test. It consists of over five-hundred questions that require a 'yes' or 'no' answer. The test was originally constructed to distinguish among several major psychotic and neurotic states. The simplicity of the questions is at once a charm and a danger. Some of the questions seem destined to produce evasive or untrue answers (Is your sex life satisfactory? Do you lie?). But the questions are easy to answer, and a 'lie scale' is built in to assess the subject's tendency to contradict and to deceive.

Studies of MMPI scores of criminals show reliable differences on several neurotic and psychopathic scales. Although these differences are significant statistically for the group, it is difficult to predict the likelihood of criminal behavior of any single subject based on his or her scores on the scales. The Gluecks ran into the same problem, though not with the MMPI, and well summarize it:

"A meaningful pattern does tend to emerge from the interweaving of separately-spun strands.... Delinquents are more extroverted, vivacious, impulsive, resentful, defiant, suspicious, and destructive. [Interestingly, in light of the common difficulty in any attempt to separate the criminal mind from the psychotic mind, these same characteristics would also describe bi-polar or paranoid mental disorders.] They are less fearful of failure or defeat than the non-delinquents, probably a recognizable description of 'successful' politicians, novelists, and the military, as well. They are less concerned about meeting conventional expectations, and are . . .less submissive to authority [perhaps a suitable description of the traits of successful artists and artisans.] They are, as a group, more socially assertive. To a greater extent than the control group, they express feelings of not being recognized or appreciated." [22]

Freud's Followers and the Criminal Personality

The psychoanalytic approach must be given credit, or blame, for the most powerful attempt to unite the legal and medical understanding of criminality into a unitary theory explaining why criminal and neurotic behavior develops in some, but not in others. In the United States Freud's theories were pushed forward by his German and British disciples who had been displaced from Europe by Nazi-Germany. Psychoanalysis changed the folk-belief of everyday life through newsprint, drama, literary theory, and film. '*All about Eve*', and, especially, '*Rebel without a Cause*,' and, later, the play '*Equus*' are examples of how psychoanalytically based drama became prevalent within the public consciousness.

Personally, Freud warned against psychoanalytic treatment of adolescents and criminals (a suggestive interpolation), but this caveat was ignored by followers.

These followers believed that psychoanalysis could be used to search-out events occurring early in life that, becoming repressed, are 'acted out' through the criminal act. Cutting through three generations of comments on psychoanalysis and crime, it is fair to summarize the central position thus: 'the need to be punished for unresolved Oedipal guilt is an important causal factor in criminal behavior.... occasionally a criminal takes such foolish risks that he appears to be trying to be caught...[the psychoanalyst sees] self-neglect, injury, and the self-defeating behavior in confinement as testimony to a need to be punished.' [23]

The Oedipal crisis occurs when the child's attentions to the mother are met with strong rebuff from the father. All of us are presumably victims of this triangular drama, and we 'work through' it in different ways and to different degrees of success. We are all 'punished' for our incestuous feelings, and from this punishment we generalize our sense of guilt and come to imitate the gender of our punishing parent. But some of us become punishers, both as parents and as criminals – for criminal behavior is a way of finding and punishing victims. Of course, these are not conscious processes, but unconscious ones.

The scenarios of crime are high drama. The actors' parts are in place: A criminal's acts will titillate us, but it turns out he too is truly just a victim of a long-repressed memory. The doctor, who patiently evokes descriptions of dreams and infantile experiences, finds a pattern through authority and intuition. Once the crucial variable is uncovered, the doctor explains the criminal act and cures the criminal by allowing her or him to understand the true cause of the heinous acts. Thereby, the criminal is permitted a new life, a life presumably of self-understanding.

The crime, then, is merely a substitute for the repressed, taboo act. We should have no trouble naming examples of this plot, for it is a staple of modern literature, film, and drama. More recent expressions of the theme have identified separation-trauma, especially separation from mother, and repressed aggression, as likely contributors. Presently, 'maladaptive parenting' and 'dysfunctional families' are much talked about. If this style of explanation is proffered, we might assume that criminal behavior is not bad. Crime is explicable, to be expected from the nature of evils applied to one's past, a natural psychological consequence of society's over-concern with sexual content, weak parenting, and a punishing penal system. The current system mistakes illness for choice, rather as if medicine punished the patient for having a fever.

Empirical Psychoanalysis

Perhaps the most complete psychoanalytic analysis of criminal personality was done by Samuel Yochelson, Stanton Samenow, and their colleagues. They performed long-term analysis of thirteen patients at a Washington D.C. hospital who had been found 'not guilty by reason of insanity.' [24] The purpose of these interviews was simple and commendable: "We wanted to know what differentiated mentally ill criminals from nonmentally ill criminals and each of these from other segments of the psychiatric population." [25] In addition to analytically oriented interviews, chromosomes were examined, along with fingerprints, palm prints, blood and steroid chemistry, brain waves, and that durable favorite, body-types. That work was supervised by the seemingly omnipresent William Sheldon.

Following the trend of improved experimental design, commendable controls were used. For example, the interviewers were unaware of the nature of the crimes associated with the individuals. Not all patients were found to be informative, some appeared to be clearly psychotic and hence inappropriate for analysis. For example, consider patients X. He could not recall his crime. He remembered that a friend, a woman with a child, rejected his sexual advances. He remembered taking the child to his home, but nothing more. In fact, he had stabbed the child, tossed her from a high window, jumped himself, and survived. Under sodium amytal, the 'truth serum' so well used by Hollywood to produce desired dramatic memories, this patient recalled pieces of the crime, but not the crime itself.

The 'successful' subjects were members of a 'treatment-group' that met for fifteen hours a week in addition to spending an hour a week with an analytically-oriented psychiatrist. These sessions cannot be described as advice-giving or coddling. They appear to have been, more or less, standard analytic sessions in which the patient/subject set the agenda and worked to solving whatever problems he wished to present, guided from time to time, by the analytic suggestions of the psychiatrist, who, himself, was gathering first-hand data on the relationship of psycho-analytic concepts and criminal behavior.

The author-analysts provide a judicious review of their findings, most of which appear to suggest that the one quality their patients had in common is a lack of feelings of guilt. This finding might seem contrary to the idea that repressed guilt is the cause of criminal activity, but the Freudian scheme of things marches our thinking backward and upside down: it is the fact that the guilt has not been acknowledged or expressed that leads to the continuation of criminal behavior. (We should remember that this group of patients is unusual, for each was judged criminally insane, a judgment often sought but rarely given.) Consideration of the thinking patterns of these patients, as expressed in the analytic sessions, was thought to give insight into the personality of the insane criminal. I here abstract some suggestions:

"The criminal is tremendously energetic . . As a child he showed more intense motor activity than the average. The attention span is short; there is a faster, more intense quality to his play... by the age of ten this activity turns to the forbidden. [26].... His thinking is rapid, continuous, and intense. [This description corresponds with the Gluecks' findings.] He attacks new tasks, such as reading. He may be all over the classroom ... the activity is not random, it makes life more interesting. As he grows older, the flow is organized into fantasies and schemes. [27].... Fear is widespread, persistent, and intense throughout life... It is so pervasive as to seem independent of experience. [28].... The fear of death is very strong, persistent, and pervasive in the criminal's mental life. He lives every day as though it were his last. The preoccupation with death is not a phenomenon only of childhood, but prevails throughout a lifetime. [29].... He perceives threats emanating from many sources.... Criminals stay away from doctors and dentists. . . .They fear being put down by people or events. [30] . . The criminal fears being reduced to 'nothing' more than he fears almost anything else: this zero state paralyzes him in a way that other fears do not. [31].... anger begins as an isolated event but spreads and spreads. Eventually, he decides that everything is worthless. [32].... criminal pride corresponds to an extremely and inflexibly high evaluation of oneself. [33].... There is sentimentality... the criminal prides himself on being a good person, one having done some nice things for people, especially for family. [34].... he wants to be a powerful person — combat hero, fireman, law enforcement agent, clergyman. His fascination is with power — fast cars, motorcycles, chemistry sets.... he is aware that dress sets him and his importance apart [35].... a continuing theme is a power thrust and a rejection of legitimate power." [36]

Observe that some, if not all, of these descriptions fit most of us, but, according to these authors, a 'deep channel runs through criminals' thinking about themselves'.

These criminals are not 'good' patients. They fail to engage themselves in the positive transference that Freudians believe necessary for analysis to progress. Increasingly frustrated, Yochelson and Samenow tell us, "It is as though a patient who desires a physical examination refuses to get un-dressed – the doctor has no access to the information he needs." [37] Their patients conceal things and are paranoid (though it seems to me that suspicion when incarcerated may be prudent). Their patients 'open' only when it serves a purpose, that is, when the truth might save but a lie do disservice. It is a matter of 'tactics'. Openness is not a way of life: openness is to be used. The criminal is often preoccupied. As someone talks to him, his mind entertains a scheme for another crime, conjures a woman, or considers ways of bailing out of trouble. He simply is not paying attention.

[38] The criminal rejects criticism of himself, but actively criticizes others. [39]

Perhaps tellingly, the criminal portrays himself as victim. Criminals view society as unfit for them, rather than regarding themselves as unfit for society. The world does not give them what they think they are entitled to. The case of participant C provides an example:

"Although his father was not in the home, the family was a stable unit. The mother, aunt, and uncle lived together and tried to use modern psychology and good judgment in rearing. Despite a stable home and many opportunities, C was constantly violating. Finally, in exasperation, his mother whipped him once. From then on, he went around telling others that he was beaten at home. Later in life, he told a psychiatrist that he had been cruelly treated as a child." [40]

The attentive reader notices that the descriptions are taking us away from the concepts of psychoanalysis. The descriptions are those of a mind tuned, and capable of retuning itself quickly, to self-interest. Guilt is instructive only because of its comparative absence. If Yochelson and Samenow's lengthy and detailed work with criminal patients, read from end to end, provides a moral, it thus: Even these authors ultimately fail to reveal something definitive about whether psychoanalytic concepts can explain such behavior.

The power of the ideas that form the basis of psychoanalysis, however, is not to be denied. Psychoanalytic concepts have become part of our way of thinking about ourselves and others. We explain, especially in the inexplicable behavior of others, as repressed memories, traumatic events, dysfunctional parenting, incest, sexual and physical abuse, for these are terms and shorthand explanations that serve to give us an understanding of our own minds and those of others. No example better shows how these ideas are represented to the public than the book and film *Rebel without a Cause.*

Rebel without a Cause: A Case Study of the Criminal Mind

Whether we should regard Robert Lindner [1905-1956] as a founder or victim of the attempt to explain criminality is questionable. While an undergraduate, Robert Lindner undertook a senior thesis at the then newly-built United State Federal Penitentiary at Lewisburg, Pennsylvania, a building of gothic design whose architecture rivaled those of the great, older universities of England and America. His book, *Rebel without a Cause*, is, perhaps, his best-known work, although it is usually confused with the 1955 film of the same name, with which it has little in common. (Book titles are not copyrightable.)

The book and film do share one important aspect, at least for our purposes. Both well illustrate the idea that crime and the criminal can be explained by the use of psychoanalytic concepts. Lindner is fond of the saying "Behind every crime lies a secret", and he repeats this observation in most of his works. *Rebel*, the book, offers a technique sufficiently dramatic to appeal to the drama desired by film and theatre, one by which the patient is induced to remember childhood events both through the 'couch' technique and through induced hypnosis. Lindner refers to the addition of hypnosis to the cure as 'hypnoanalytic.' Rebel, the book, reports the transcript of forty-six hours of hypnoanalysis undertaken with a patient/criminal named Harold. Harold is approximately sixteen years of age and appears to have advanced from petty theft to armed robbery, a series that led to his commitment to the prison.

Harold is not merely a delinquent or criminal undergoing punishment or penitence: Lindner regards him as a 'psychopath', a term describing behavior today more commonly called 'sociopathic'. The book attempts to discover the secret that underlies psychopathic persons, for however difficult it may be to define 'psychopath' accurately, as the saying goes, you know one when you meet one. Lindner is even keen enough to reminds us that a 'psychopath' is only a psychopath within a particular culture, for the psychopath is identified by his or her denying of societal values. In a capitalist state, unremorseful theft may be psychopathic. In a Hindu state, drinking alcohol without remorse might qualify.

Still, the "psychopath is a rebel without a cause, an agitator without a slogan, a revolutionary without an agenda: in other words, his or her rebelliousness is aimed to achieve goals satisfactory to himself alone. He or she is incapable of exertions for the sake of others. All efforts, hidden under whatever guise, represent investments designed to satisfy immediate wishes and desires." [41] The psychopath is infantile in that she or he cannot put aside erotic gratification and cannot wait for prestige. The wish, the need, is for immediate gratification. Just as the infant cries when hungry and accepts no substitute, so the psychopath is unable to put aside needs and wants for a more appropriate time. The super-ego—that mental filter that gives us guilt and encourages us not to act—- is weakly constructed, thereby suggesting to the psychoanalyst the likely importance of the oedipal situation from which the superego arises.

The hour-long meetings between Harold and Lindner are recorded by a transcriber in another room who hears the proceedings by means of a microphone hidden in the couch. Lindner justifies this secretive technique by pointing out that Harold, afterward, suggests a recording of the events – a justification that would not be considered ethical today. Harold's narratives follow an orthodox analytic line, and Lindner helps us, on highly selected occasions, by drawing our attention to evidences of resistance and transference that might escape all but the working analyst. We learn, at first, of Harold's childhood, his associates, relations with parents, then we learn more detail as the sessions continue. We discover that since early childhood, perhaps infancy, Harold had a disease of the eye, most probably of the retina. That led to his fluttering his eyelids, which in turn resulted in his being teased severely. We also learn about Harold's life in the reformatory, especially about the developing relationship with two friends. Here is a sample of Harold's narrative, accompanied by an unusually long comment by Lindner:

> [Harold] "Once in a while my mother would call me a blind bat or something like that in Polish, or my sister sometimes would say something. I used to hang around with my cousin Riggs and he'd call me names like Squint. I don't know why I hung around with him, I disliked him so. I never committed any crimes with him because I hate him so much."

[Lindner, to the reader] The obverse. Only with those whom I love can I commit crimes enhances the significance of this amazing statement. As we shall see, Riggs was a father-substitute and thus hated. What Harold means is that the forbidden act is the forbidden [the criminal act is the sexual taboo] and— for persons like himself — can only be performed with mother surrogates as a substitute means of gratifying the hidden wish. [42]

Later we find that Harold rejected a friend because Harold suspected he was making sexual advances. After recalling a dream about the friend, Harold says:

"As for the dream, I don't know... I was really angry with him. He hasn't said anything since... I'll probably just forget about it. When I start thinking about a person, the first thing I think about is what do they think about my eyes".

Lindner hears this as the onset of transference, that critical relationship that develops between analyst and patient by which the patient comes to understand his true relationship with mother and father. Transference is the heart of the cure: left incomplete or unexplained by the therapist the patient leaves worse off than before. [40] (We may be reminded of the events surrounding Victor Race and the Marquis de Puységur. In addition to initiating suggestive hypnosis as a medical cure, between them they may have offered one of the first descriptions of the transference process.

Still later, now the twenty-second hour, Harold develops his memories of infantile sexual interest, events by which we approach the specifics of his gender and superego development through the Oedipal period.

[Harold] "The first time I saw my sister naked, it reminded me of a man without a penis. I didn't see a penis. It was something strange. I never asked about it. When I went with Lila [this, years later, during childhood] I liked to play with her breasts but she would get so hot she'd reach inside my pants and play with my peter; so I'd lay down and have intercourse with her. I disliked seeing her skinny legs. I thought it looked like a man with his penis and testes cut off. [We learn later that Harold always keeps his eyes closed during sexual intimacy and that he pulls the bed-clothing over his head at night so he can't see]

[Lindner] "Who would be most likely to cut them off?"

[Harold] "Why the father or the mother or the doctor"

[Lindner, to the reader] Note the order of primacy in which potential castrators are placed. [43]

[Harold] "I don't think I ever saw my mother completely naked. I never saw my father completely naked."

Upon later reflection, Harold modifies these recollections:

[Harold] "As far back as I can remember I didn't like my father. I would never speak to him unless it was necessary. When I was about twelve I got into trouble by breaking into a store. When I was sent home, my father didn't say anything to me."

[Lindner] "If you saw your parents being intimate, how would it appear to you?"

[Howard] 'Well, it appeared [note the tense] that my father was hurting my mother. Maybe I did see my father and mother do that. I can't recall. It must have been way back before I can remember. It might seem vulgar, brutal, filthy, dirty or what not." [44]

[Howard] "I guess I did feel a little resentment against my father for touching my mother. . . . When I'd hear them in bed, hear them talk and him coaxing my mother I hated listening to it. I'd put the covers over my head and try to shut out everything: sometimes I'd recite nursery rhymes. . . . I still sleep with the covers over my head. I hated to listen to it. I didn't want to. I wanted to be away from there. I'd pull the covers over my head..."

[Howard] "... My father, he used to threaten me with the dog. . . he used to say things about my eye and curse me out. My sister was a tomboy and he would say things about cutting off my penis and giving it to her. [45]

As the sessions continue, some under hypnosis, some not, Harold recalls more and more happenings that carry the threat of castration and he finds himself remembering more memories of himself an at an earlier age. Lindner pushes:

[Lindner] "Now Harold, you are in your mother's arms. She is holding you. You have the measles. There are spots on your hands. Are your eyes open?"

231

[Harold] "Yes — no — no. They are blinking. . . . everything is bright. Bright. I can't look at it for a long time. I have to blink my eyes.

Here, Lindner tells us that perspiration accumulated on Harold's face and he lifts his hands to pull his clothing away from his body.

[Lindner] "I am going to ask you a question. Listen carefully. Why did you first start to blink your eyes? . . . . Now you are very small, very small. This is very long ago, when you were in the cradle."

[Harold] "I am in the cradle, right next to my father's bed. I see him. I see — I can see it way up. My mother —I see my father on top of her. I'm in the cradle — and I see him. I — it is early in the morning — not very dark — not light — and — I'm in — the cradle. My mother's nightgown is up over her hips, and she is on her back — and I see his — I see my mother. The light is coming in."

When Lindner asks Harold how old he was at the time of this memory, he replies, "one year of age." [46]

Again, during the thirty-fifth hour:

[Howard] "I saw him in bed — one morning — I woke up — My mother looked — naked to me; my father had his underwear or something on him. . . . When my father looked over at me I saw his genitals so I got more afraid that he'd be coming over to me and hurting me with his genitals. I — I — my father was hurting her. [47]

232

When the Oedipal story becomes repetitive, Harold announces that he is guilty of killing a man, this after he has commented that he often felt like killing his father, especially when he thought of the sexual relationship between the parents. Presumably, now that Harold recognizes the elements of the Oedipal stories, he feels free to confess the murder. Lindner now has before him a major ethical problem, the resolution of which Lindner held for a footnote at the end of the book, and which I shall hold for a bit longer. [48]

In the meantime, Lindner assists Harold in interpreting how his present criminal actions have their causes in previously unrevealed feelings.

[Lindner] "There are many reasons why people do things; some are rational, some are disguised to appear rational. You used a weapon which, for you, is a symbolic representation of the penis, and which you had stolen from your father. You actually used the weapon most suited to your working through of your conflict. You used, on the representative of your father [the man Harold believes he has killed] the substitute of the weapon which he had taken from you, and which you now stole from him."

[Lindner] "A child sees something that is forbidden. . . You, the child, become prey to the feelings of guilt, and these are later intensified by feelings of inadequacy. You run away by closing your eyes . . . That thread runs through your whole life. You are running away . . . your whole life is a struggle with him and a running away from him and your inadequate self. "

[Lindner] "And the fear of castration that you have always had; it's shown very well by many incidents. You slept with your aunt [an incident that happened when he was twelve] and you awakened to find her hand on your penis. She wanted to steal it from you! And you dislike her. One time you slept with a man who was a boarder, and he also slept with his hand on your penis . . . that was when you said a man broke into the house and was trying to steal something."

[Harold] "I was afraid of his hands. When he said he'd sic the dog on me to bite my penis off I was afraid of him then. I was small, but he looked so big . . . I was afraid of him, his big arms, his big powerful muscles." [49]

Freudian psychoanalysis holds that in order to be fully cured, transference must happen, but then be resolved, or overcome. Harold's success at this was probably mixed. Near conclusion of the relationship Lindner writes:

[Lindner, to the reader] Harold's behavior during the forepart of this session demonstrates his ambivalent attitude toward the writer and the therapy. On the one hand, to hide his disappointment caused him by his correct anticipation that treatment was drawing rapidly to a close, he took occasion indirectly to chastise the writer and to minimize the therapeutic benefits. On the other hand, like all patients, he fought tigerishly against surrounding the neurotic so-called 'secondary gain' which had until now provided him with reasonable protection and the excuse for his behavior. [50]

All analysis has only a pragmatic ending-point: there is neither world nor time enough to finish. Such is clearly true for the treatment of the criminal psychopath, as it seems forever unlikely that society would care to shoulder the cost, even were such treatment proved effective. Lindner has left us with a legacy, however; this of how analytic ideas might be applied to understanding the mind of the psychopathic criminal. Of which Harold's mind, to be sure, is but a single example.

Robert Lindner, having educated us by his account of the treatment of Harold, would not have us marching away humming a cheerful tune about how society might handle the criminal. He concludes *Rebel without a Cause* thus:

[Linder, to the reader] "We have had in this volume a striking illustration of the truth of William A. White's remark that behind every criminal deed lies a secret. But more important, we have glimpsed the utter futility, the sheer waste, of confining individuals in barred and turreted zoos for humans without attempting to recover such secrets . . . Harold plundered and almost killed in response to those ungovernable needs which came flaring up from the deepest, remotest shafts of his being. Had he not undertaken analysis, all the trade-training, all the attentions of penal personnel would have been wasted on him; and like every other psychopath who leaves prison he would have been released again to the community as the same predatory beast who entered — with this exception; that his conflicts would have been driven more deeply and his hostility aggravated by a system that flatters itself that it is doing other than substituting psychological for physical brutality." [51]

Douglas Keith Candland

Stone Walls and Men

The book *Stone Walls and Men,* published two years later, in 1946, represents Lindner's straightforward attempt to use psycho-analytic ideas to locate the motivations of the criminal. The first parts of the book contain long autobiographies written by convicted criminals. Throughout these are, to Lindner's ear, ample evidence that serious crime has as its cause the substitution of criminal acts for a forbidden, unconsciously desired one.

Consider Linder's account of the case of C. R. (sometimes, described as only "C"), here abstracted:

C. R's crime has deep roots. He was born in a large Northern city, a healthy and robust youngster, the second of two children. His parents were incompatible and were held together only by their children and their religion. But their marriage was intolerable to both, and when C was three, they separated. C's mother, an intensely religious woman, neurotically-inclined, opinionated, possessed of strong prejudices, somewhat lacking in warmth, decided that she could not maintain the children at home. When C was four, both children were placed in an orphanage. They were destined to spend upward of sixteen years in similar circumstances.

[There] he was beaten, made to do extravagant penances for minor disciplinary infractions, and was exposed to the brutality, the sadism in fact, of the frustrated personnel of these monstrous institutions. The climate of the homes was outstanding for lack of warmth and affection. . . He became indifferent to the sufferings of others and came to delight in giving pain. He became shrewd and cunning in his efforts to cast blame on others, to obtain favor. His sexual life was oriented toward the abnormal by the situation in which he lived.

Finally, C.R. was 'finished' with his schooling and allowed to go home. He and his brother went to live with the mother. The mother, however, persisted in treating C and his brother as children, although by now they had at least the physiques of men. She dictated every phase of their lives, maintained a tight control over the family purse. He stole sums of money from her, once broke open her trunk and lifted and then pawned her wedding ring.

The sexual energies which had for so long been dammed within him were vying for outlet. He made tentative advances toward girls in his circle but these were turned aside. When he resorted to prostitutes he was impotent. Only through masturbation could he quiet his desires, and then merely temporarily and with much guilt.

On the day of the murder, C. found himself to be lacking in funds. He knew that his mother kept cash and her wedding ring in a trunk in her bedroom. He decided to break open this trunk and to do this he made his way to a neighbor who loaned him a hammer. He encountered a girl who was selling some article of apparel from door-to door. She inquired whether his mother was at home. He replied that she was and offered to direct her to his apartment. She preceded him into the apartment and he indicated the general direction of the bedroom. As they passed the kitchen he lifted the borrowed hammer from a shelf and struck her. She fell to the floor. Her moaning excited him: he ran to the kitchen, grabbed an ice-pick and drove it into her body many times.

In the analysis of C. R. it developed that the crime he had committed was in reality a displacement of an act he had been preparing to commit all his life. The intended act was the murder of his mother.

Now the clue to behavior resides in the Oedipus situation. He had been deprived of his father at a time when this infantile conflict was coming to a head. In this case, there was no opportunity to work through the distressing complex, and he was perennially in the grip of the profound infantile attachment to the mother. This attachment must be viewed as one which could express itself mainly aggressively, and which was insoluble because a satisfactory paternal figure was missing....

Not only did C. have a powerful infantile love for his mother, but he also hated her for abandoning him....

C's dreams and thoughts as well as his symbolic acts — such as breaking into his mother's trunk and stealing her wedding ring - -- were all indicative of his basal want. He desired to possess his mother, and according to his notions as they had been generated by brutalizing and heartless experiences, this could be done only by force. In his masturbatory fantasies he imagined all types and varieties of such an incestuous act, but they all entailed gruesome and gory accompaniments such as were enacted in the murder. The homosexuality in which he had engaged found him always the aggressor, and he could be potent in heterosexuality only with mother-substitutes. But because of his rigid religious training, C could not really assuage his intimate desires through the possession of the mother: he could not conspicuously conceive of committing incest. The desire had to be displaced. So it was that the unfortunate girl met her end. [52]

Many of the other stories tell similar tales, supporting Linder's hypotheses. He concludes his book by telling us that

Not education of parents alone, not the psycho-eugenics of mating alone, not slum clearance alone, not the psychotherapy of the individual criminal alone, not the reorganization of social institutions alone, not any separate phase or aspect will prevent crime, but all of these together and at once. [53]

The reader may pause at the notion of 'psycho-eugenics of mating'. Our contemporary view does not, in general, prompt us to think of how we might restrict matings to do away with the genes that cause criminality. But Lindner was writing during the years of World War II. The governments of at least three of the combatants had authorized and practiced such restrictions, as well as harsher ones. To augment our collection of such mental fossils, we turn to our final core idea. Once more we will see the ideas of variation, human progress, and physiognomy intertwined. We shall also see that the core ideas we have explored are not merely curiosities---taken seriously, they can become lethal.

Chapter 8

The Mental Fossil: Human Progress through Eugenics

Conspicuous remnants of the fossil record of eugenics are to be found between the first publication, in 1869, of Frances Galton's book *Hereditary Genius* and the onset of World War II, in 1939. They are to be found in three Western countries, the United Kingdom, Germany, and the United States, in which planned 'eugenics' sought to remove the 'unfit' as reproducing members of society. In *Hereditary Genius* and in his other works regarding peoples' physical and mental attributes, Galton showed that the genes of one generation were correlated with the genes of the next; for example, that people of the economic upper-classes produced offspring of like economic potential. Galton's blessing, to mix a metaphor, gave the movement not merely a helping hand, but a powerful shove. In like fashion, recognition of the importance and implication of Gregor Mendel's demonstration of recessive and dominant genes (around 1900) convinced some that human genetics was able to answer some of the riddles that Darwin asked but could not answer.

Intelligence and Feeblemindedness

That people differ in mental abilities was not a discovery of Galton's, but his analysis of how such variation might be measured provided a means for psychology to invent and embrace the concept called 'intelligence'. In Paris, while Galton was nearing the late part of his career in England, Alfred Binet [1857-1911] was working out a method for measuring variations in 'intelligence' with the politically progressive idea that people should be educated and trained according to their inherited ability.

By the application of the statistical techniques invented by Galton, Karl Pearson, and Roland Fisher came the ability to sort people statistically by their intelligence and potential for achievement. The 'testing movement' itself spawned measurement of personality traits, workers' abilities, and the putative potential to do advanced work in academic fields. SATs, GREs, MedCats, all these owe their existence and nature to Galton's ideas. Curiously, the original purpose of intelligence testing became reversed: at its onset the idea was to determine the ability of folk so each could be educated or trained to the best of his (not usually her) potential. The 'slum child,' noted the American developer of mental tests, Lightner Whitmer, can be released from that environment and given opportunity, if only we can assist the kind and degree of genetic potential so that society can seize this child and provide the requisite education and training.

Again we see that a core idea which began as a liberating and progressive one came to be restrictive. As the IQ tests were expected to promote equality, or at least a more equal-footing among people, they became a useful tools in denying such equality. Test results came to be understood as a fixed, relatively invariant measure of ability, one that classified the individual and set limits as to her or his likely attainments. Intelligence --testing became one more method which assisted the favored classes by buying education and keeping other classes in check.

Douglas Keith Candland

The Range of Eugenics

Attention to putative heredity has been of constant interest in any society for which written records exist. In fact, many important finds of physical archeology are nothing more than records of a given person's ancestry. From time to time, human societies grow weary of such passive tracing of heredity and consciously take it upon themselves to actively alter the process of genetic re-assortment. We may categorize the ways in which this is done as follows:

1. Eliminating the unfit. If a society can decide on how the unfit are to be defined and recognized, it may remove them from the society either through taking responsibility for their death (for example, the removal of the insane, the retarded, the homosexual, the Jewish, the Gypsy). A society may also elect to remove the unfit by prohibiting their reproduction by (a) sterilization, castration, vasectomization, or chemical destruction of aspects of the reproductive system or (b) removing such persons to segregated places where they may not reproduce. Both latter systems have been practiced in the United States.

2. Limiting the gene pool: The most common method to control a society's genetic makeup, if among the least successful, is by a society to restrict immigration into it. If it is assumed that others, but not oneself, carry unwanted genes, then by prohibiting folk with genes unlike your own from meeting those with which you share genes, mating is prevented and the frequency of the unwanted genes is thought to be diminished. Immigration laws passed in the first years of the twentieth century in the United States provided limits on persons from other countries. Northern Europeans were allowed access to the

United States in far greater numbers than southern Europeans, while Asians were disallowed and the number from Africa and Caribbean countries was near zero. Whatever reason may be advanced for the creation of rules favoring the immigration of one culture over another, one effect is to alter the gene pool of both.

3. Controlling Marriage: At times, society has believed that diseases thought to be of a genetic origin (psychosis, certain neurological and endocrinological diseases that are lethal, or requiring long-term care by others) might be eliminated by regulating marriage, thereby presumably regulating procreation.

Many US states had laws regarding miscegenation, often providing a fine for any clergy who performed marriages between 'races'. The definition of 'race' was usually based on the degree of 'Caucasian blood', at other times and places, on the presumed amount of Christian, Jewish, or Islamic blood. Some states had laws barring marriage between 'feebleminded and feebleminded' or between feebleminded and normal. Feeblemindedness was established by the recommendation of a physician or by a test of intelligence. Most states prohibit marriage between related persons, although the degree of relatedness allowed differs wildly from state to state. Note that these laws are based on different notions of how genes function: intermarriage between 'races' was thought to weaken one race or the other; feeblemindedness, no matter how it was manifested, was thought to be caused by a single, malformed, gene; and marriage between, say, cousins, was thought to increase the chance of the appearance of fatal or unwanted genes in offspring.

Figure 8.1: Assigned brides and grooms. The brides were to become mothers of special babies fathered by the soldiers before they went to war. From Kaupen-Haas, 1989. Permission requested from Greno Publishing.

4. Encouraging the production of offspring: A political body may decide that marriages among certain 'fits' are to be encouraged; or that women are to be encouraged to become impregnated by men deemed to be genetically appropriate. By 1934, for example, Germany was offering young women thought to possess desired genes the pleasure of childbirth in specialized hospitals where high-quality medicine and nursing was intended to decrease mortality of both infant and mother. Those women interested were impregnated by appropriate males, often by men shortly to enter the military.

An alternative, offering state money directly or through tax benefits to the children or parents of the children possessing the wanted genes appears both in the policies of the United States and in the United Kingdom. These schemes are not intended to increase the population alone, as was done in the European settlement of the American west, or to 'assist' low income, large families, but to encourage the reproduction of certain genes, obviously, if tacitly, at the expense of other, non-chosen, genes. It may be pertinent to comment that this notion at present appears to be a component of the United States tax code, although, no doubt, other justifications are given for it.

Leonard Darwin, Charles Darwin son, in the years following his presidency of the (British) Eugenics Education Society put forth various schemes, including one to tax bachelors. [1, 2] As they were not contributing to the future of their race, it was argued, they might be expected to help the government assist those people who undertook responsibility for continuing society by reproducing in quality and quantity. There were also plans to tax people according to the number of children they produced or simply taxing those who failed to bear children. There was also talk of a system much like our current one in order to encouraging population growth by offering tax credits to those who 'over reproduce'.

5. Discouraging the production of offspring. The eugenic societies feared that it was those people best able to care for children (or the 'race') who, because of education, money, and presumably fine genes, would use birth control techniques, while those less fit would breed only too happily, unencumbered by such devices. One solution was (and is) for the former to assist the later by

circulating the devices along with information on their use. The giving of such advice is not limited to individuals or various societies within a culture: countries, including the United States, have been known to wish to be helpful to other countries by offering, for example, free vasectomies or, at other times, frowning on certain forms of birth control. Several countries attempt to reduce the growth of their populations by limiting the number of offspring a couple may produce. It is rumored that these laws lead to infanticide, especially of females.

Many cultures, or significant parts of them, have disapproved of techniques used to reduce reproduction. Something of this idea was exhibited in England in 1876 by Annie Besant, who appeared earlier both through her attendance at the World's Congress of Ideas at the Chicago Fair [Chapters3] and through her association with Rudolph Steiner and the anthroposophists [Chapter 4]. Author Peter Washington tells the tale succinctly:

"In 1876 a Bristol bookseller was arrested for selling obscene literature in the shape of The Fruits of Philosophy, a misleadingly entitled tract about birth control. The book had been reprinted several times in Britain and America since its first appearance in 1833, but the bookseller had since inserted some helpful diagrams which were thought to push a dubious text over the border of decency into illegality. The original publisher of the stock was prosecuted and fined."

[Charles] Bradlaugh and [Annie] Besant became involved in the case when they republished The Fruits of Philosophy after the original trial. They soon found themselves in the dock [on trial] in 1877, where they made gargantuan speeches in their own defense, to no avail. [3]

6. Segregation of the 'unfit'. An alternative method, one most commonly used in the United States, is to construct facilities for those identified as unfit in the hope that physical segregation will remove or at least limit their rate of reproduction. As the 'feebleminded' are the most easily identifiable, and those most in need of assistance while, along with the insane, the least unable to reject the assistance offered, these folk were often placed in same-sex homes or institutions in which education and training could be offered. Not incidentally, the inmates' reproductive strategies could be monitored as well.

The Sterilization Solution

There has been, of late, a tendency to equate all ideas of human genetic control and engineering with the Nazi elimination of the unfit. This judgment, however, ignores the evidence that Germany was but one among many countries interested in arranging human genes for the purpose of human betterment.

The eugenics movement was most visible in the United States in the campaign to sterilize the unfit from around 1899 to the present. In Great Britain, although sterilization laws were approved by Parliament on two of the required three readings, the third and essential reading never came, perhaps because of the simultaneous onset of World War I. Sterilization laws were also approved in Canada [1928], Denmark [1929], and then Finland, Sweden, Norway, and Iceland, to cite western countries, and, in the east, Japan. [4]. The absence of the names of southern European countries is due to the objections of the Roman Catholic Church for whom sterilization, like the availability of birth control devices, demeaned the individual. Sterilization was both a 'Progressive' and Protestant cause in the west.

In the United States, sterilization began before it was legalized. After a Dr. Sharp vasectomized young male inmates in Indiana in 1899 with the stated hope of eliminating 'excessive' masturbation, sterilization laws were passed by the states of Connecticut, Washington, California [1909], New Jersey, Iowa [1911] Nevada, New York [1912] North Dakota, Michigan, Kansas and Wisconsin [1913] Nebraska [1915] and Oregon, South Dakota, and New Hampshire [1917]. [5] State legislatures did not design such laws on their own: they were prompted and heavily lobbied by groups who identified themselves as progressive. Like all political processes, the pressures and counter-currents were complex, but the pressures were successful. California's sterilization laws were approved by the progressive governor Hiram Johnson.

At first, the sterilization laws were applied specifically to the feebleminded. These persons were chosen because the inheritance of IQ was believed to be a genetically-simple affair and because of the power and publicity regarding a series of 'studies' purporting to show that families of low intelligence reproduced the unfit in larger numbers than the fit. Attempts to sterilize the criminal and the insane met with less favor. By the 1930s, society in the United States and Great Britain had shifted its consciousness, accepting the view that the feebleminded, criminal, and psychotic could be treated and aided to become a law-abiding, producing member of society. There was no need to remove the genes, as environment could ameliorate the behavioral difficulties produced by low intelligence.

The number of persons sterilized is reported by Mark Haller, an historian of the American eugenics movement, to be the following:

"By the end of 1931, twenty-four years after the first law, somewhat more than 12,145 operations had been performed under the laws, 7,548 of these in California alone. At that time sterilization of the insane outnumbered sterilization of the feebleminded about two to one, while sterilization of epileptics, criminals, and others was negligible. By the end of 1958, the total sterilizations under the laws had risen to 60,926. Although California still led with 20,011, such southern states as North Carolina, Georgia, and Virginia were performing more sterilizations relative to their populations. By then the number of persons sterilized for feeblemindedness had slightly topped the number sterilized for insanity, and about three out of five of those sterilized had been women." [6,7]

By 1958, the number of reported sterilizations in California was reported as numbering 13 during the last two years. The United States was not singular in these attempts. In March, 1972, a Canadian Blackfoot Indian, Leilani Muir, was forcibly sterilized as part of the long-standing plan of the Province of Alberta to remove the reproductive potential of mental defectives, these including girls as young as 14, already infertile Downs Syndrome boys, children with cerebral palsy, illiterate immigrants from Eastern Europe, and delinquent youths from bad homes. [8] The total of such persons appears to have been 2,800. Though the Roman Catholic Church had argued strenuously against the bill passed in 1928, scientists from the major Canadian universities, especially McGill University, had lobbied successfully in its favor. When Ms. Muir sued, the defense mounted the argument that the sterilization fit the mores and scientific understandings of the time.

What stalled the sterilization movement in America was not, a change in our understanding of genetics or psychoology. Nor was it an appreciation of the complexities of human heredity, which might have been gained from reading the work appearing from the Galton laboratory in London (about which, more later). Instead it was due to the American courts' interpretation of the United States Constitution and, especially, that part known as the Bill of Rights. But, as we will see later, such a view was long in coming.

Cold Spring Harbor and the National Registry of Genes

In the United States, foundations supported the plan to establish a national registry of each person's genetic 'stock'. Such was a major task of the Eugenics Record Office at Cold Spring Harbor, Long Island, then directed by C. B. Davenport. Davenport had previously been among those folks based in Washington D.C. engaged in measuring aspects of peoples of different cultures, especially American Indians. Supported by the Carnegie Corporation of Washington, The Eugenics Record Office took as its purpose the collecting of information on the presumed heredity of Americans by means of forms filled out by children as a school assignment. These required school children to have their parents complete 'genetic forms' or to have staff officers take records during personal visits. Such knowledge, it was believed, would go a long way to helping people plan their reproductive futures and ultimately toward more intelligent and the therefore advanced societies. The goal was to develop a national registry for Americans, from which any US citizen could evaluate his genetic ancestry and presumably inquire about those of others.

The Bulletins from the Office well illustrate the potential use to which such data might be put. A 1911 bulletin features *The Heredity of Feeble-mindedness,* containing some of Henry Goddard's investigations of families said to be 'beneficial' and 'unbeneficial.' It was reprinted, perhaps informatively so, from the American Breeders Magazine. In 1915, Davenport published *The Feebly Inhibited: Violent Temper and its Inheritance* and. in 1923, *Body-Build; its Development and Inheritance.* Many articles were also written by H. H. Laughlin, secretary of the group, whose documentation of his understanding of human heredity represented a spirited and seemingly tireless concern. Mark Haller, a historian of the American Eugenics movement, writes that Laughlin was devoid of humor. [9] This must be so: how else can we account for the title of Laughlin's 1914 work *Report of the Committee to Study and to Report on the Best Practical Means of Cutting Off the Defective Germ-Plasm in the American Population?*

At Cold Spring Harbor, the view that the inherence of undesired traits was a simple matter of removing a gene persevered under Davenport and Laughlin's guidance. Under their auspices, Madison Grant published on the dangers of immigration to the 'American' gene pool. Of course, by 'American', Grant meant Nordic or North European. Grant's books were well-received, although not by the academic establishment, and they well-exemplify the single-gene notion. Grant, Davenport's college classmate, was a well-regarded member of society whose public works included saving the California redwoods, the presidency of the New York Zoological Society, and a trusteeship of the American Museum of Natural History. At the Museum, he had worked with Henry Osborn, whose influence extended over the 'hiring' of Ota Benga [Chapter 4] and the murals on the development of races that graced the Museum's halls. [Chapter 3] In short, the leadership of the eugenics movement in the United States was, may one say, inbred? Althogh both the Carnegie Corporation of Washington and Cold Spring Harbor function in our times, both now have different mandates and goals.

Yerkes and the Genetics of Personality

In order to establish the personality traits that might have a genetic locus, the young, but already distinguished, comparative psychologist Robert Yerkes, working with a colleague at what is now East Stroudsburg State University of Pennsylvania, in 1913 prepared a self-help document offered to the individual "to help you to understand yourself and to become a useful and happy member of society . . ." [10]

The recipient of the questionnaire, a pamphlet of 24 pages, was asked to record information to be sent to Yerkes and his colleagues from which notations could be made in a master file about genetics and personality. These data, it was thought, would show that the kinds and qualities of behavior followed genetic lines. The Introduction to the booklet clarifies Yerkes's and LaRue's view:

"The authors have discovered, through their experience as teachers, that a study of the ancestry, development, and present constitution of the self is an extremely profitable task for most students. . . . The purpose of this study is threefold: first, to help you to understand yourself and to become a useful and happy member of society; second, to help you to understand and sympathize with other persons, especially children, and to further their development; third to arouse your interest in the facts of heredity, of environmental influence, and in the significance of the applied sciences of eugenics and euthenics.

Be sure to arrange for personal conferences with the instructor and with some of your relatives in order that you may obtain advice and assistance in gathering information.

The object to be studied is the self. We shall study it (1) as a product or expression of heredity; (2) as a developing, reaching mechanism (a going machine); (3) as a conscious and self-conscious willing being; and (4) as a member of social groups." [11]

There follows a table in which one may insert information regarding grandparents and siblings, as well as parents and oneself: among the categories to be rated are ability in vocal music, drawing or coloring, literary composition, calculating, remembering, sensory defects, temperament (slow, intermediate, nervous quick), use of hands (handiness), birthmarks, hare lip, and abnormal fingers or toes.

The similar 1914 *Ancestral History of the Self* by Yerkes and LaRue includes notes on one's family traits (bodily, mental, moral, and social) through great grandparents, information about oneself as a child and adolescent, and similar information about oneself at the time of one's recording the information. The record portion of the pamphlet is labeled *Record of Family Traits* with the information to be sent to the Eugenics Records Office at Cold Spring Harbor. The heading offers up the ennobled title of the Carnegie Institution of Washington, Department of Experimental Evolution.

"Temperament and character. Make as truthful a portrait of your temperament and character as you can. Point out what you deem defects and discuss possible ways of remedying them. Indicate, in contrast with your actual self, your ideal. Who have been your heroes during adolescence? Why? Who is now your ideal?

In the following list underscore once the terms that fairly well describe you as an adolescent; underscore twice those that describe you very accurately. Here are the selections made by the unknown student seemingly around the year 1914:

Respectful, disrespectful; responsive, unresponsive; capricious, steady; prompt, procrastinating; resourceful, helpless; gentle, violent; objective-minded, subjective-minded; graceful, awkward; purposeful, desultory; thorough, superficial; orderly, disorderly; optimistic, pessimistic; contented, querulous; originative, imitative; careful, careless; deliberate, rash; industrious, indolent; practical, Dreamy; persistent, fickle; visionary, matter of fact; sanguine, melancholic; critical, suggestible; romantic, unromantic; systematic, unmethodical; erotic, cold; excitable, stolid; emotional, lethargic" [Here the original test-taker lost interest in self-description.]" [12]

SYMBOLS FOR NORMAL & VALUABLE TRAITS & QUALITIES	SYMBOLS FOR DEFECTS

Figure 8.2 shows the information requested.

Now that scientists have allegedly maped all of the human genes (at the least, they have fully mapped a person's genes) the idea of a master-list showing the genetic origins of every individual is possible. The idea of a National Registry is poised for success.

The Vineland Training Center

The Vineland Training Center was, then, a home for the 'feebleminded'. Having found that many of the institutionalized feebleminded also had feebleminded relatives and ancestors, a huge leap of genetic faith was made by saltating to the belief that feeblemindedness was a matter of direct genetic inheritance, and studies were begun to so demonstrate. The Vineland Center published results of their collections, often in the Eugenics Record Office publications. Henry H. Goddard worked there, and published much, though in later years he repented that

that he had oversimplified the issue of the genetic cause of feeblemindedness.

Thirty of the state legislatures in the United States by 1931 had passed laws permitting sterilization and castration, sometimes for the feebleminded, sometimes for the insane, and one for chicken-thieves and car-thieves. These laws were based, chiefly, on the understanding by the eugenicists of the way by which disease and behavior are transmitted genetically. Figure 8.3, taken from Goddard's work, displays this viewpoint. The view that alcoholism and feeble-mindedness are inherited directly, presumably by a single, seemingly dominant gene, appears evident: Stillbirth, infant death (of unknown cause) and miscarriage are, by implication, thought to behave in a like fashion.

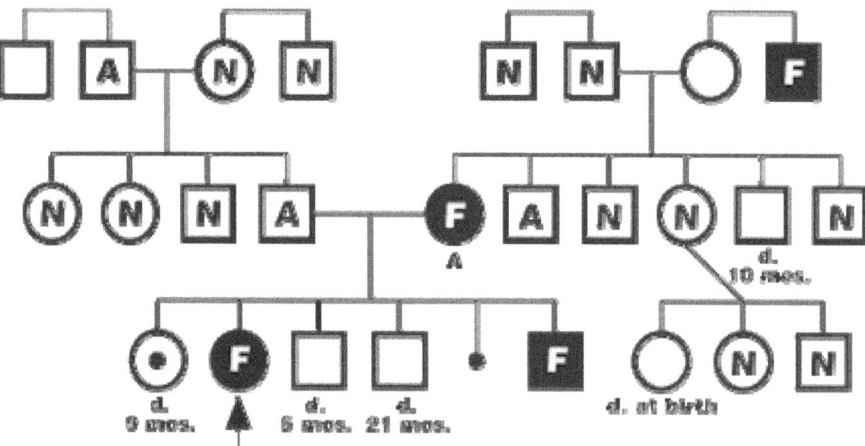

Figure 8.3: from Goddard, 1911, p. 3 "Chart II shows a combination of alcoholism and mental defect in the ancestry of the parents, resulting in alcoholism on the one side and direct feeble-mindedness with alcoholism on the other. The offspring of these two individuals are all defective — one stillborn, two that died young, one miscarriage, and two feeble-minded. Chart III [here presented] is instructive, in that it seems to show the effect of a combination of

alcoholism and mental defect in the father; when mother's family is good — herself and sisters being normal. The result of this woman's marriage with a feeble-minded alcoholic man is five feeble-minded children, five that died in infancy, two others that died before their condition could be determined, and one normal child. Apparently a clear case of transmission through the father."

Buck v Bell Revisited

In the United States, sterilization was legalized and, as we have seen, often enough performed. The legal case of Buck v Bell (Figure 8.4) illustrates the power of law, and I here rephrase the account of this case offered by author Mark Haller:

"Born to a feebleminded and immoral mother, Carrie [Buck] had been adopted at the age of four . . . There she attended school until the sixth grade and then, unable to continue further, did housework under strict supervision. But she proved unmanageable and was soon pregnant. In January 1924, as a result, she, like her mother before her, was committed to the State Colony for Epileptics and the Feebleminded, where her child was born. In the fall she was chosen for the first sterilization under the Virginia law passed that same year, and her case began its course toward the Supreme Court of the United States." [13]

When the United States Supreme Court heard arguments in 1925 in Buck vs. Bell, the matter of Carrie Buck's mental age, and that of her mother and child, was introduced. According to H. H. Laughlin, previously described, her ancestors were a "shiftless, ignorant, and worthless class of anti-social whites of the South". Other testimony offered that feeblemindedness was a single, recessive, Mendelian trait. If so, I comment, it follows that some of her children may well have been of 'normal' or 'greater than normal' intelligence, although eugenicists rarely appreciated the point.)

When similar cases had been heard before several other state supreme courts, the courts were disturbed by the vagueness of the laws, the difficulty in defining the mental conditions required for sterilization, and, significantly, the fact that the laws permitting sterilization appeared to apply to some groups (the institutionalized) but not to others. Some state supreme courts were further bothered by the lack of due process, chiefly the question of whether the person had received information regarding the surgery proposed. However, in this case, the order for sterilization was upheld by both the Circuit Court and the Virginia Supreme Court. Eventually, in the spring of 1927, the United States Supreme Court, held that sterilization fell within the police power of the state. In his famous decision, Justice Oliver Justice Holmes wrote:

"It is better for all the world, if instead of waiting for their imbecility, society can prevent those who are manifestly unfit from continuing their kind. The principle that sustains compulsory vaccination is broad enough to cover cutting the Fallopian tubes. [14]"

Discharged from the home, although surely sterilized without her knowledge or permission, Carrie later married and appears to have lead a reasonable life, although one without children. The account of her life, and especially of the arguments presented at the trial, is a powerful exhibition of the awesome power of persistant ideas to enforce belief, and an equally powerful warning of what may happen when we accept without investigation the most recent cultural dressing, whether we are a retarded Carrie Buck or a justice of the Supreme Court of the United States. [15,16]

Figure 8.4: Carrie Buck and husband. In a Virginia home for the retarded, she had been sterilized without her knowledge in order to eliminate the probability of 'defective genes' being transmitted. Permission requested from Smith and Nelson and New Horizon Press, 1989.

Douglas Keith Candland

"A Race of Shy Men"

Charles Darwin and Francis Galton's contributions to eugenics are, in Darwin's case, seemingly intentionally ambiguous, and in Galton's, absolutely direct. There is no little irony to the fact that Charles Darwin's direct genetic contribution to society, his children, promoted the eugenic movement, as did many educated and progressive persons of their time. The Darwins and Galtons did not, of course, invent 'human eugenics'; they did, however, provide the scientific authority on which the modern movements were based. There have always been eugenic practices, these varying from the idea of a celibate priesthood, to immigration laws, and legal and religiously-based views regarding marriage and procreation, each of which shape the genetics of following generations.

Galton's contribution to the modern eugenics cause, however, was far more practical than his discovery of ways of measuring variability [Chapter 3]. On May 16, 1904, he, with the statistician Karl Pearson, took the message of human eugenics to the Sociological Society that "just as animals could be bred to eliminate some traits and enhance others, so human society must take care of its future". Galton endowed a fellowship in eugenics the following year and, upon his death in 1911, his estate provided funds for an endowed chair in eugenics at the University of London. Karl Pearson was named to the post. [17]

Pearson once inquired, just how much eugenic control was really wanted. Do the English want, he asked, probably rhetorically, 'a race of shy men?' Pearson's appointment led to the serious development of the study of human heredity which would later be continued by the statistician R. A. Fisher. Most of the work is statistical, rather than biochemical. More or less at the same time, an organization known as the Moral Education League redeveloped itself into the Eugenics Education Society. From this group was to come the speeches, publications, and pressure on government, enrollment of academics and schoolteachers, and testimony from persons with well-known names. Among these were (Major) Leonard Darwin, fourth son of Charles, who was president of the society and Darwin's 2nd son Francis, the botanist.

The period just before World War I is noticeable for the rapid evolution of eugenics groups. The Cold Spring Harbor group in the United States was begun by the Carnegie Foundation of Washington in 1910; the Internationale Gesellschaft fur Rassenhygiene (Race-hygiene) began in Munich the same year (Galton agreed to be the honorary president); and the first International Congress of Eugenics was held in London in 1912, the formal picture of which features Leonard Darwin in his role as president. Figure 8.5 shows the officers of the group of 800 who attended from various countries with President Darwin in the foreground. In the back is Mr. Kellogg, US representative and scion of the Kellogg cereal group. The group was to meet again under the hosting of Henry Fairfield Osborn, frequent visitor to these pages in Chapters 2 and 3, and director of the American Museum of Natural History. The meetings were held at the Museum in New York in 1921 and again in 1932.

Figure 8.5. Officers of the Society at The First International Congress of Eugenics. London, 1912. [The next two meetings were to be held in New York City at the American Museum of Natural History.] Sitting, left, is President Leonard Darwin. Standing, seventh from left, is Vernon Kellogg of the USA and sitting next to Darwin is Sybil Gotto of England. From Koch, 1996, p. 43.

Galton's working to spread the eugenics movement has meant that his name and accomplishments supported all aspects and branches. In fact, as Searle [23] makes evident, Galton was disinterested in 'negative eugenics', this the elimination of the unfit, the sort of eugenics that was successfully promoted in the United States and Germany, but instead in 'positive eugenics', this the shaping of future genetic combinations by encouraging appropriate matings. Leonard Darwin, too was also chiefly interested in promoting the matings of folk with good genes, folk who were productive members of society.

Eugenics Today

Around the world today, eugenics is in full swing. Hardly a year goes by without a significant genocide happening somewhere in the world, and the average American seems barely to notice, if at all. In our country, the movement has ended, at least in its overt form. Policies and trends that are eugenic, by effect, if not intent, however flourish even in the United States.

In our time, our understanding of evolution has been enhanced by the recognition, prompted by E. O. Wilson, R. Trivers, and others now described as 'sociobiologists', that mating — which re-organizes of genes — is not random. Human beings have in place laws, ethics, religions, customs, and social mores that shift the probabilities of individuals' selecting one rather than another as partners for mating. Hence, culture, religion, law, race, etc. are not, as is so often thought, merely byproducts of evolution. Rather, they determine the course of evolution by altering the probability of various sets of genes combining.

Religions and laws clearly differ regarding the number of husbands and or wives allowed, the number of offspring thought reasonable, and acceptance of suicide. There are still subcultures that bestow honor or shame upon those who fail to engage mates or to reproduce, that discriminate against those who have previously reproduce unhealthy children, that put at a disadvantage those who differ in religion and, most conspicuously, 'race'. There is still the tendency to ensure that wards of the state, be they criminals or mental patients, are segregated by sex and excluded from extensive physical interaction with visitors. Further, many anti-psychotic medicines now in use, either taken voluntarily or by court-order, remove sexual desire.

Douglas Keith Candland

The core ideas of phrenology ad constitutional psychology are to be found in the develoment of eugenics, a fact offering both an intellectual and a moral lesson. The intellectual lesson is the theme of this book's attempt toward an arcehopsychology: that core ideas appear and re-appear, chanhing their dress with the time and culture, but retaining their core. The moral lesson is that ideas have consequences. The irony is that each of the core ideas considered, from animal magnetism to eugenics, were considered in their time to be progressive discoveries whose application would lead to human betterment. We dare not presume to understand others and ourselves without locating the lasting core of any idea from which we decide to use our political and social skills in order to change society by controlling behavior.

Chapter 9

Revisiting Archeopsychology

Fossils are, by definition, dead things, only suggestions of what core ideas once were rather than explanations of what they have become. Mental fossils, to the contrary, are both dead and alive: dead in the sense that they seem to us to be buried, yet alive in the sense that with effort we may resurrect them by removing their disguise and our seeing that they are everywhere about our ideas regarding our minds and behavior. To say they are dead is not to say that they are unstudiable or unable to teach us. What may be learned from it is that though we think we make decisions in a seemingly rational fashion, they are often, upon the deepest sort of reflection, biased by powerful, unseen, redressed, notions, of which the of ideas described in this book are examples.

When the organizers of the Chicago Fair of 1893 offered "Human Progress" as the fair's theme, they expressed optimism that human evolution was moving forward. Distinguishing 'forward' from 'backward' evolution is a chancy prediction in the best of times, unless we are fortunate enough to know how evolution is to end. The theme, nonetheless, was represented in the displays of living human cultures. Of course, the theme 'human progress' was also meant, one assumes, to draw attention to the exhibits of human achievement in invention of handwork and tools. These ends were served by the use of electricity to power machines, and the stunning display of weaponry that would, in that century to come, the 20th, extend from mechanical to the nuclear and, in the case of nascient anthropology, neurology, and psychology, the use of photography to make both permanent and memorable other peoples, other cultures, and ourselves. We take for granted electricity and photography today, but these were marvelous spirits in these earlier times, for both involved how unseen power could translate one world in to another. The splendid armaments at the Fair were impressive, for the Krupp guns of 1893 will be put to devastating use in short order. The mayor's idea that the Fair will promote peace is now known to be hollow. The turning on of the electric lights that each evenings turned night into day created awe. But we readers from the 21st century are not surprised by unseen power: we expect radio, television, and telephone waves to pass through our house and bodies.

And what of those almost insignificant displays of neurology, anthropology, and psychology? Their use of the photograph and techniques of measurement would enhance during the 20th century human thought and understanding of oneself and one another's cultures. Just as 'animal magnetism' would lead to hypnosis, and hypnosis to the idea of the unconscious, phrenology would encourage the ideas of measuring the body's 'constitution' and the predictive value of mental tests. Mental measurement and constitutional psychology would combine to encourage in the prediction of the criminal. Studies of physiognomic variations, when paired with the notions of progress, would, in turn, make eugenics acceptable to progressive folk, the government, and in many cases, the average citizen. In like fashion, the concepts of and from mental testing provided a scientific basis for the various practices of eugenics.

The narratives used in this book led us along a certain path, yet into detours and unexpected adventures. Such is the nature of exploration, whether through uncharted sea, unmapped land, or an archeological expedition. Along our stroll, we found remnants of ideas. When examined with care, these mental fossil remains appear to speak of a basic structure that is modified this way and that. At varying times and places, the structure appears different, yet when we remove the changes we see a stable core. The precise cause of each change baffles us, yet we are certain that if we knew more of the working of mental evolution, we could understand powerful principles of psychology. For now, we can but bask in knowing that we can generate hypotheses that offer novel and effective ways of understanding the evolution our minds, of encouraging us to restructure our understanding of ourselves. Most such hypotheses will prove wrong; but the metaphor is the right one.

What we have found in the way of samples of mental fossils are, of course, not a systematically collected sample, but a collection of what history chose to keep, however randomly, and what we choose to see. Had the records of the World's Fair not been meticulously kept, had libraries not retained books on phrenology when public opinion gave it up; had, say, German become a lost language: had English not replaced German as the language of late 20th century scholarship, these fossil remnants may be unknown to us. Let us set our sample of mental fossils on the table so we may touch, turn, and examine them. Then let us explore how we can employ an archeopsychological method to aid our understanding of ourselves.

Douglas Keith Candland

Human Progress

The first narrative, embracing the world of late nineteenth century understandings of evolution of the mind and body, was illustrated and implemented at the Chicago Fair. There displays of the varieties of human groups and their achievements gave authority to the everyday notion of a 'mental' and cultural ladder of human beings. Phrases such as 'Organic Evolution' and the 'Great Chain of Being' [Chapter 2] embodied the belief that evolution was the mechanism by which human beings of differing physical and mental levels are produced.

Darwin, Huxley, and Wallace's hypotheses provided the opportunity to make scientific (according to some) and God-given (according to others) the categories by which humankind was then understood. At the time of the Chicago Fair in 1893, the races and cultures were 'objectively' divided according to their level of development, such levels seemingly having been based on German, British, and American standards being topmost. The tools and products of these cultures were displayed in halls built especially for that purpose, while Asian and black cultures were presented along the Plaisance in situ, so to speak, in their 'natural state' [Chapters 2]. The fairgoer could not miss the distinction in appreciation between European products and Asian and Black cultural oddities.

These displays could not have been honored and accepted unless they fit the idea of Human Progress being ordained and occurring through natural selection and variation. However expressed through time, the idea that humankind is differentially developed is long-standing and perhaps the most powerful of themes in regard to its effect on cultures and human judgment: ancient Egyptians are compared to Romans, written languages to oral ones, Greeks to Barbarians, Northerners opposed to Southerners, city folk compared to rural folk, men compared to women. On we go.

Of course, only when the core idea of human progress is accepted, can human progress be controlled. We saw the effects of such an attitude through eugenics [Chapter 8], but we also saw its effects in attempts to aid criminals, mental patients, lovers, and craftspeople; though we think some of those attempts misguided, we surely recognize them as noble, to some extent. Numerous tests and measures given to our children are purported to encourage the advancement of civilization, and their methods differ only slightly from those we would dismiss.

Spiritism

A second fossil collection on our table are the ideas of Steiner, Freud, and Jung who, following in Fechner's tradition, illustrate the power of the core idea of spiritism [Chapter 3], this a belief in unseeable worlds. Fechner and Steiner thought these worlds to be ultimately material; Freud and Jung thought them to be entirely mental. No difference: these are but examples of the theme of 'other worlds.' Fechner's idea of measuring how we might translate perceptions between the real and the spiritual world led him to invent the psychophysical methods, methods sometimes credited as the beginning of experimental psychology.

To this day, many of a more 'scientific' mindset claim to reject spiritism in all its forms. Yet we have discovered in our digging a connection between spiritism and the mainstays of modern science. A belief in electricity, magnetism, natural selection, and photography all require faith in unseen powers. So too do psychoanalysis and cognitive psychology. As physics and astronomy become ever-more sophisticated, their unseen powers become even more abstract, yet still there is great faith in their reality.

Variation

Modern psychology and modern social science are based upon, and pride themselves on, measurement of variation. More significantly, they use such measurement to deduce and induce conclusions. The development of measurement of variation has moved, in recent times, from measurement focused on means and averages to that which focuses on deviations. Numbers, of course, are, as Plato understood, pure; it is when they are assigned to 'things' that their meanings confuse us. We can add 5 and 3; we cannot add five apples and three oranges unless we change the concept to 'fruit'. We can measure my IQ as 100 and yours as 200, but not thereby conclude that you are twice as intelligent. The number 90 is a different number, depending on whether it represents intelligence, fruit, a grade, or the number on an athlete's jersey.

The development of the use of numbers to categorize human beings is reflected in the development of techniques used in phrenology, as well as body-typing, and mental testing, to making assumptions regarding genotypes, thereby to eugenics. Phrenology [Chapter 4], a technique emphasizing the measurement of variation, is a manifestation of the notion that variations in the human body are predictive of mental aspects. Phrenology began as a method to test a reasonable hypothesis; namely, that body and mind are one, so that measurement of one is revelatory of the other. From its origins in the intellectual, scientific, and philosophical cultures of Scotland, England, France, and Germany, the methods of phrenology spread elsewhere in Europe, then decayed. The idea flourished in North America, where its scientific value shifted to the practical (and commercial) by measuring the skulls of the insane and feebleminded in the hope of establishing ways of predicting such unfortunate social variations. Later, the techniques migrate to reside exclusively in commercial markets.

Measurement of human variation is essential to constitutional psychology, the measurement of body-type (as opposed to only the skull's shape) and its relation to temperament and, later, to criminality. A straight line can be drawn to the use of measurement of variation to determine, as in the practice of eugenics, differences in mental capacity. But applying numbers to variations tells us nothing. The danger is when we assume the meaning of variation. As phrenology discovered how to succeed commercially, the correlations between physical shape and ability became expressed achieved merely by claiming them. For example, a measurer specifies an area of the cranium as measuring amativeness, and then proceeds to measure this area from a number of people. Some have 'more', some 'less.' Averages and deviations of amativeness may now be computed. But, unless these measures are correlated with some behavioral measure of whatever amativeness is, we have measurement without meaning. Even with seemingly significant correlations, we have no evidence of causality. More dangerous, we have

Douglas Keith Candland

seemingly verified the measurability of a concept, when, in fact, we have done nothing but established a tautology.

Physiognomy

The relationship between temperament and body type has as long a human history as any recorded idea. Although its strength wanes in different cultures at different times, and while it is now in the West abated, we look to the recent past, indeed, to a generation but twice removed from our own, to find its raging strength. There is truly, one suspects, a relation between body type and temperament, for evolving ideas cannot be forever built on false grounds. At some level of human understanding, temperament and physical structure correlate. But all things correlate: number of hairs on the head and length of toe surely correlate at some level, but correlation does not offer causes, nor much accuracy in prediction.

Baleful evidence of the truth of this warning can be found in the instructive studies of the physiques, temperaments, delinquency, and lives of the young described first in the 1920s then in the 1950s. How rare it is for social science to offer comparative data made on the same people thirty years apart. If these studies, restated and reviewed in Chapters 5 and 6, are not precise measures of the reliability of the measures, they nonetheless stand as instructive. For these young men, World War II altered their lives. Most of them spent years in one of the armed forces, most often the Army. Some were killed, some died in middle age; some married and fathered children; most worked in industry, some had careers of a professional nature. It is not surprising to learn that both their bodies and their lives were altered in unpredictable ways, surely because of the war.

The demise of phrenology came, in the United States, not because it was seen to be unscientific or undemocratic, but because its core was reclothed into the measurement of other variations, particularly the patterns produced by filling in bubbles on questionnaires. It is senseless to laugh at phrenology, or to point to its unscientific-ness as if this were cause for its dismissal. The important fact about phrenology for us is that its validity was accepted by people living throughout the nineteenth century.

For the lay populace, it should be noted, the belief that body reflects mind may have changed little in content over many years, though its overtness has certainly lessened. Again, we look at our youth: In casual advice giving, how much are the career choices and relationships they are encouraged to pursue dependant on their appearance? When trying to answer this question, we find that we are all phrenologists and body-typists. We human beings do make assumptions about people we meet based on our perception of their general physique or aspects of their bodies, eyes, noses, hair, and more. We prejudge. The issue is not whether we do so, or even whether we ought to do so, but how we do so and what we make of our assumptions. Consider, as a phrenologist might, the shape of the human toes or the shape of the earlobe. There are, I am told, two chief kinds of toe patterns, called the Dutch and the English. The difference depends on whether the slant from big toe to the little toe is linear or whether the toe next to the big one is larger, thereby upsetting the linearity. It is probable that few people in our times notice this distinction, much less that they ascribe characteristics of personality to this seemingly slight variation in human anatomy. Earlobes, too, take one of three main types, but we are unlikely to learn that we have been given or denied a job because of the type we were given. Maybe not, but the same neutrality is clearly not true of skin color.

Statistics and Prophecy

Statistics are helpful: they help us to find unseen relationships, to assess reliability, to understand interactions. But statistics are themselves mute and mechanical: they react only to the information we give them. For this reason, they are also capable of masking information if we fail to ask the 'right' question. Nor can they answer questions we forget to ask of them. Any experiment, any collection of data, is, of course, limited by the factors chosen: if 'neuroticism' or 'orality' do not distinguish our groups, perhaps the terms are too broad, and we might find differences if we were able to limit our inquiry to some more refined measures. Alternatively, perhaps the terms are too narrow, and we might find differences if we were to enlarge the definitions of these terms. In short the finding of no difference between two groups does not mean that the groups are not different: it may mean that the factors selected are not sufficiently powerful or sufficiently precise to allow us to detect differences. It may be that whatever differences exist are interactions of these variables which are often hard to detect. For example, we might find that only extreme insecurity coupled with extreme extroversion yields criminal behavior, while extreme security coupled with extreme introversion does not.

The prophetic tasks of the investigator are twofold: First, before beginning the collection of data, they must identify the significant factors. This is an ironic task, to be sure - if one knew the appropriate variables, there would be no need to perform the experiment. Second, after they have gathered and analyzed the data, they must read the statistical tea-leaves for suggestions as to where more meaningful differences and interactions might be sought. Usually, for reasons of time, money, and perhaps because of the influence of core ideas, social science rarely makes it to the difficult, but crucial, second step.

To realize that a large, expensive, collection of data is the beginning, not the end, is an admission that social science rarely makes. Perhaps, given the power of folk-psychology and dislike of long-term commitments in the funding of social science, it is a road rarely offered. Perhaps the strength of the ideas being investigated clouds our judgment. Perhaps social science rests too easily once its preconceptions are seemingly confirmed.

The Danger of Mental Fossils

I have, from time to time, both described and hinted at the potential and current dangers of not recognizing the influence of the powerful ideas here represented as mental fossils. By narrating aspects our recent intellectual history, I have hoped to lead the reader to self-recognition. These mental fossils demonstrate ideas that are not trivial or neutral: they are the figurative roots of human policy toward one another. As filters of our minds, they determine how we interpret the world around us. Identifying them is more than helpful introspection: it is the propaedeutic method of any science of the mind.

A major source of the authority and power of a fossil idea is that it appears to offer an explanation for current belief: we pluck the needed one, change it to fit the contemporary circumstance, and honor it as an intellectual or (as is most honored in our times) scientific discovery. The evolving idea acquires a power over us, because we find it easier and more convenient to examine each appearance rather than searching hard for the underlying core. Which of our current celebrated ideas are but another fashionable dress of an ancient idea, one that can deceive us into approving things that may be dangerous? This question cannot be answered, nor prophylactics prepared, if we concentrate only on the present. But if we examine enough fossils, we may find that they have in common a backbone, or an ear, or an optic tectum. The study of change through evolution, whether of genes or ideas, is our hope for not mistaking today's variation as ultimate.

Evolving ideas are nefarious, because, through their manipulation of science and other authority, they can encourage the belief that common sense is basically correct and that our beliefs are intuitively accurate. Both merely reinforce whatever prejudices we bring to our understanding of the world. Later phrenology, such as that of Fowler, appears to have assumed the relationship between skull and talents. No attempt was made to investigate the assumption. The core idea became truth, unquestionable and unquestioned. What was at first an unproven hypothesis becomes a new 'fact', with little, if any, work needed inbetween.

To see the ability of ideas to manipulate science, we need only return to our examination of delinquent youth. We ask the data to tell us which aspects account for the differences between delinquents and nondelinquents. But if we stop the investigation when we have merely re-identified the same variable with which we started our work, we have accomplished nothing and worse, we may have done damage by fueling prejudice. That is, by ceasing our study at this point of correspondence, we merely confirm the prejudice with which we designed the study.

We must also be weary that our predictions do not become self-fulfilling, for the effects of this upon science, and society, is devastating. Does the labeling of an individual with information about the probability of an ulcer, or a crime, an estimate of amativeness, or the prediction of a successful profession or occupation limit or enhance the probability of success, illness, or criminal behavior? What other unforeseen consequences might an emphasis on such predictions have? Can we resist the urge to claim that a correlation between body-type and temperament indicates that personality, character, and temperament are determined genetically? Even if these predictions might be used to assist individuals and make for a more just society, could we have avoided the shift from phrenology and constitutional psychology to eugenics?

It is not surprising that we so often fall into these traps, as society encourages investigations that it understands and which search for finds with which it already agrees. The investigator that follows this path finds rewards - a certain celebrity status and funds for continuing researches being examples. One wonders how different the world of natural and social sciences would be if investigators published under a code number to be decoded, the name revealed, only after death

Douglas Keith Candland

A Task for Archeopsychology

We begin to suspect that social and behavioral science might be but investigations about ideas made through the filter of other ideas. What turns the investigator's attention to this idea or that is, at the 'practical' level, not necessarily logical (itself an idea) but a reflection of the common sense then prevalent in the concerns of society. If the suspicion warrants deeper consideration, we might consider whether social science is in unwitting acceptance of powerful ideas, rather than in investigation of them. Perhaps social science might prosper by turning its attention toward investigating the core of ideas and their essences rather than assuming that each variation that influences it represents a separate expression of some form of truth.

It will not be an easy task to identify, much less to catalog, core ideas. Our task must be historical, as we are unlikely to be able to see around the ideas that occupy the thinking our times. Alas, the history available to us is, of course, selective. The task is made more difficult, at least for now, because we have no history of doing so, not even attempts shown to be misguided, and none shown to be successful. But it is not enough to merely call for arms: we must have examples to serve the role that 'critical experiments' play in the sciences, which are as important for their method as for their results. The digs I have performed, their results herein arranged and presented, seek to serve such a function.

A psychology that fails to escape from the orbit of its own ideas is destined to do nothing more than repeat them (to paraphrase George Santayana). Alas, research continues unabated that, with the glow of scientific truth, meretriciously fails to add to our knowledge of ourselves. Nay, much effort is expended to merely add to the darkened filter that already overlays the task of understanding ourselves and other living beings. We can at least illuminate our way by first studying the beliefs and hypotheses that guide our questionings; the themes by which we organize our perceptions and knowledge. We must uncover the ideas that lead us to search.

That is the first task of a true psychology of the mind.

BIBLIOGRAPHY

Abrahamsen, D. "Crime and the Human Mind," New York: Columbia, 1944.

Ackernecht, E. H. and Vallois, H. V. Franz Joseph Gall, Inventor of Phrenology and his Collectionî [translated from the French by Claire St Lèo.] Wisconsin Studies in Medical History, Number 1. Madison, Wisconsin: Medical School, Madison, Wisconsin. 1956.

Ahern, G. "Sun at Midnight, The Rudolf Steiner Movement and the Western Esoteric Tradition," Wellingborough, Northamptonshire, UK: The Aquarian Press, 1984.

Alexander, R. "The Biology of Moral Systems," Hawthorne, New York, USA: A de Gruyter, 1987.

Ambler, A. T. and Banta, M. (Eds.) "The Invention of Photography and Its Impact on Learning; Photographs from Harvard University and Radcliffe College and from the Collection of Harrison D. Horblit," Cambridge, Massachusetts, USA: Harvard, 1989.

—-- American Academy of Political and Social Science. "Race Improvement in the United States," Philadelphia, 1909, Annals 34, Number 1.

—- American Eugenics Society, "The Development of Eugenic Policies; Scientific Backgrounds for a New Orientation of Eugenics," New York, 1937.

—- American Neurological Association. "Eugenical Sterilization; a Reorientation of the Problem," New York: Macmillan, 1936.

Appelbaum, S. "A Photographic Record: Photos from the Collections of the Avery Library of Columbia University and the Chicago Historical Society, The Chicago World's Fair of 1893," New York: Dover, 1980.

—- Archiv fur Rassen-und Gesellschaftsbiologie, einmschliesslich Rassen - und Gessellschaftshygiene," Jahrgenge 1904-1911, Leipzig, Teubner.

Badger, R. "The Great American Fair, the World's Columbian Exposition and American Culture," Hall: Chicago, 1979.

Bain, A. "On the Study of Character, Including an Estimate of Phrenology," London: Parker, 1861.

Bakan, D. "Sigmund Freud and the Jewish Mystical Tradition," Princeton, NJ, USA: Princeton, 1958, 1959.

Baldwin, J. M. "The Story of the Mind", New York: D. Appleton, 1898, 1901. Banta, M and Hinsley, C. M. "From Site to Sight, Anthropology, Photography, and the Power of Imagery," Peabody Museum Press: Cambridge, Massachusetts, USA: distributed by Harvard Press, 1986.

Barkow, J. H., Cosmides, L. and Tooby, J. "The Adapted Mind: Evolutionary Psychology and the Generation of Culture," New York: Oxford Press, 1992.

Bauer, K. H. "Rassenhygiene, ihre Biologischen Grundlagen," Leipzig, Quelle and Meyer, 1926.

Berlin, I. "The Hedgehog and the Fox," Oxford, UK: Blackwell, 1980.

Bernhein, H. "Suggestive Therapeutics, a Treatise on the Nature and Uses of Hypnotism," [C. A. Herter, translator.] New York and London, Putnam's sons, 1880. The date of publication given in the book is incorrect, as the work refers to events through 1890.

Bertillon, A. "Signalled Instructions," Translated from the French; translator not identified. Chicago: Werner, 1896.

Besant, A. W. "The Ancient Wisdom: an Outline of Theosophical Teachings," [tenth edition] Adyar, Madras, India: Theosophical Publishing House, 1977.

Blais, H. "Les Tendances Eugenistes au Canada," Montreal: L'Institut Familial, 1942.

Bolotin, N. and Laing, C. "The Chicago World's Fair of 1893, The World's Columbian Exposition," National Trust for Historic Preservation: Preservation Press, 1992.

Box, J. F. "R. A. Fisher, the Life of a Scientist," New York: Wiley, 1978.

Bradford, P.V. and Blume, H. "Ota, The Pygmy in the Zoo," New York: St Martin's, 1992.

Bratlinger, E. "Sterilization of People with Mental Disabilities, Issues, Perspectives, and Cases," Westport, Connecticut, USA, Auburn House, 1994.

Bridges, W. "Gathering of Animals, An Unconventional History of the New York Zoological Society," New York: Harper and Row, 1974.

Brigham, C. C. "A Study of American Intelligence," [Foreword by Robert Yerkes] Princeton, New Jersey, USA: Princeton Press, 1923.

Bromberg, W. "Crime and the Mind, a Psychiatric Analysis of Crime and Punishment," New York: Macmillan, 1965.

Brooke, J. L. "The Refiner's Fire, The Making of Mormon Cosmology, 1644-1844," Cambridge, UK: Cambridge Press, 1994.

Brown, J. K. "Contesting Images, Photography and the World's Columbian Exposition," Tucson and London: University of Arizona Press, 1994.

Burg, D. F. "Chicago's White City of 1893," Lexington: Kentucky Press, 1976.
Burkert, W. "Creation of the Sacred: Tracks of Biology in Early Religions," Cambridge: Massachusetts, USA: Harvard, 1996.

Burks, B, Jensen, D. W., and Terman, L. M. "The Promise of Youth: Follow-up Studies of a Thousand Gifted Children," Volume iv in "Genetic Studies of Genius," Stanford, California, USA: Stanford, 1925.

Burt, C. L. "The Young Delinquent," New York: Appleton, 1925.

— California, State Joint Committee on Defectives, "Surveys in Mental Deviation in Prisons, Public Schools and Orphanages in California," Sacramento, California, USA: State printing office, 1918.

Candland, D. K., "Feral Children and Clever Animals," New York: Oxford, 1993.

Carpenter, P. F. and Totah, P. [Eds.] "The San Francisco Fair, Treasure Island: 1939-1940," San Francisco, California, USA: Scotwall, 1989.

Cattell, R. B . "The Fight for our National Intelligence," [Introduction by Major L. Darwin]. London: King and Son, 1937.

Chesterton, G. K. "Eugenics and other Evils," New York: Cassell, 1922.

Choris, L. "Voyage Pittoresque au tour du Monde, avec des Portraits de Sauvages d'Amerique, D'Asie, D'Afrique, et des iles du Grand Ocean; des Paysages, des Vues Maritimes, et Plusieurs Objects d'histoire Naturelle; Accompagne de Descriptions par M. Le Baron Cuvier, et M. A de Chamisso, et d'obsrevations sur les Cranes Humains, par M. Le Docteur Gall," Paris: Didot, 1822.

Churchland, P. M. "The Engine of Reason, the Seat of the Soul; A Philosophical Journey into the Brain," Cambridge, Massachusetts, USA: Massachusetts Institute of Technology Press, 1995.

Churchland, P. M., "A Neurocomputational Perspective; The Nature of Mind and the Structure of Science,"

Cambridge, Massachusetts, USA: Massachusetts Institute of Technology Press, 1989.

Churchland, P and Churchland, P. See also Macauley, R. N. [Ed.] "The Churchlands and their Critics," Oxford, UK: Blackwell, 1996.

—--"Columbische Weltausstelling im Chicago," Leipzig: Gustav Frinche, 1893.

Combe, G. "The Constitution of Man, Considered in Relation to External Objects," Hartford, Connecticut, USA: Silas Andrus & son, 1849. (The original Edinburgh, Scotland edition is 1825.)

Combe, G. "Elements of Phrenology," fourth edition: Edinburgh: MacLachlan and Stewart, and John Anderson, 1836.

Combe, G. and Combe, A. "On the Functions of the Cerebellum by Drs Gall, Vimont, and Broussais, also Answers Objections urged Against Phrenology by Drs Roget, Rudolph, Prichard, and Tiedemann," "On the Functions..." is a translation of Gall, 1838. Edinburgh: MacLachlan and Stewart, 1838.

Conklin, E. G. "The Mechanisms of Heredity, " Science, 1908, 89, p. 27.
Crabtree, A. "Animal Magnetism, Early Hypnotism, and Psychical Research, 1766-1925, An Annotated Bibliography," White Plains: New York, USA: Kraus, 1988.

Crabtree, A. "From Mesmer to Freud, Magnetic Sleep and the Roots of Psychological Healing," New Haven, Connecticut, USA: Yale, 1993.

Crane, R. N. "Marriage Laws and Statutory Experiments in Eugenics in the United States," London: Eugenics Education Society, 1910.

Dahlstrom, W. G. "Personality Systematics and the Problem of Types," General Learning Press, 250 James St., Morristown, New Jersey, USA: 1972.

Daley, S. "Kazza Kamma Journal: Endangered Bushmen Find Hope in a Game Park," Article, New York Times, International Edition. Thursday, February 18, 1996.

Dana, C. F. "The Modern Views of Heredity with the Study of a Frequently Inherited Psychosis." Medical Record, 1910, 77, p. 245.

Danielson, F. H. "The Hill Folk: Report on a Rural Community of Hereditary Defectives," Cold Spring Harbor, New York, USA: New Era Printing, 1912.

Darwin, C. for "Charles Darwin's Beagle Diary," see Keynes, R. D. (Ed). Darwin, C. "On the Origin of Species by Means of Natural Selection; or, the Preservation of Favored Races in the Struggle for Life," London: Murray, 1859. Six revised editions through 1872. The final edition is the one usually reprinted. The later revisions omit the "On" used in the title of the earlier editions.

Darwin, C. "The Descent of Man, Selection in Relation to Sex," (Two Volumes). London: John Murray, 1871.

Darwin, C. "The Expression of the Emotions in Man and Animals," London: John Murray, 1872. Reprinted in paperback, Chicago: Chicago Press, 1965.

Darwin, L. "Eugenics and National Economy, An Appeal," Presidential Address by Major Leonard Darwin, June 12, 1913. London: Eugenics Education Society, 1913.

Darwin, L. "The Need for Eugenic Reform," New York: Appleton, 1926.

Darwin, L. "What is Eugenics?" [Second Edition] London: Watts, 1929.

Daum, M, and Deppe, H-U. "Zwangssterilisation in Frankfurt am Main 1933-1945," Frankfurt and New York: Campus, 1991.

Davenport, C. B. "Eugenics, the Science of Human Improvement by Better Breeding," New York: Holt, 1910.

Davenport, C. B. "State Laws Limiting Marriage Selection," Bulletin 9, Eugenics Record Office. Cold Spring Harbor, Long Island, New York, USA, June 1913.

Davenport, C. B. "The Feebly Inhibited; Nomadism, or the Wandering Impulses," Washington DC, USA: Carnegie Institution of Washington, 1915.

Davenport, C. B. "Army Anthropology, Based on Observations Made on Draft Recruits, 1917-1918 and on Veterans at Demobilization, 1919," Washington DC, USA: Government Printing Office, 1921.

Davenport, C. B. "Body-Build and its Inheritance," Washington DC, USA: The Carnegie Institution of Washington, Publication 329, 1923.

Davenport, C. B. "Body Builds: Its Development and Inheritance," Bulletin 24, Eugenics Record Office. Cold Spring Harbor, Long Island, New York, USA, February, 1925.

Davies, J. D. "Phrenology, Fad and Science," New Haven, Connecticut, USA: Yale Press, 1955.

Davies, R. "The Conscience of a Writer, " In "A Voice from the Attic" New York: Knopf, 1960.

Dawkins, R. "The Selfish Gene," [New Edition]. Oxford, UK and New York, USA: Oxford, 1989 [first published, 1976]

Deleuze, J. P. F. "Practical Instruction in Animal Magnetism," Translated from the French by T. C. Hartshorn. [Revised Edition]. New York: Fowler and Wells, 1879.

Dennett, D. C. "Darwin's Dangerous Idea, Evolution and the Meanings of Life," New York: Simon and Schuster, 1995.

Dennett, D. C. "Appraising Grace, What Evolutionary Good is God," Review of Burkert. New York: The Sciences, January/February 1997, pp. 39-44.

Desmond, A. "Huxley; from Devil's Disciple to Evolution's High Priest," Reading, Massachusetts: Addison-Wesley, 1994-1997.

Diamond, J. "Guns, Germs, and Steel, the Fates of Human Societies," New York: Norton, 1997.

Dight, C. F. "History of the Early Stages of the Organized Eugenics Movement for Human Betterment in Minnesota," Minneapolis, Minnesota: Minnesota Eugenics Society, 1935.

Dinage, R. "Annie Besant," New York: Penguin, 1986.

Donald, M. "Origins of the Modern Mind, Three Stages in the Evolution of Culture and Cognition," Cambridge, Massachusetts, USA: Harvard Press, 1991.

Dugdale, R. L., "The Jukes: A Study in Crime, Pauperism, and Heredity," several editions: New York: Putnam's, 1910. See also, as cited by Haller, p. 201, "A Record and Study of the Relations of Crime, Pauperism and Disease," Prison Association of New York: Thirty-first Annual Report, 1875.

Dunlap, K. "Personal Beauty and Racial Betterment," St Louis, Missouri, USA: Mosby, 1920.

Ellenberger, H. F. "The Discovery of the Unconscious, the History and Evolution of Dynamic Psychiatry," New York: Basic, 1970.

Ellis, H. "The Task of Social Hygiene," Boston: Houghton Mifflin, 1912.

—--"Encyclopedia of Occultism and Parapsychology," Three Volumes, Second Edition. Shephard, L. (Ed.). Gale: Detroit, 1984.

—--Eugenics Congress, International [the Third, 1932], American Museum of Natural History, "A Decade of Progress in Eugenics; Scientific Papers of the Third International Congress of Eugenics," Baltimore, MD, USA: Williams and Wilkins, 1934.

—--Eugenics Society, London. "Symposium of the 19th Annual Meeting: London, 1982," London: Academic, 1983.

Fechner, G. T. "Elemente der Psychophysik, " Leipzig: Breitkopf und Hertel, 1860.

Fechner, G. T. [written as Dr. Mises] "Kleine Schriften," Leipzig: Breitkopf und Hertel, 1875.

Fechner, G. T. "In Sachen der Psycho-physik," Leipzig: Breitkopf und Hertel, 1877.

Fechner, G. T. "Revision der Haptpunkte der Psychophysik," Leipzig: Breitkopf, 1882.

Fechner, G. T. "Uber die Seelenfrage. Ein Gangdurch die Sichtbare Welt, um die Unsichtbare zu Finden," Hamburg: Voss, 1907.

Fechner, G. T. "Zend Avesta: oder, uber die Dinge de Himmels und des Jenseits vom Standpunkt der Naturbetrachting," Leipzig: Voss, 1922.

Fechner, G. T. "Nanna, Oder uber das Seelenleben der Planzen," Leipzig: Voss, 1921. (Ed. K. Lasswitz).

Fenner, E. "Zwangssterilisation im Nationalsozialismus, Zur Rolle der Hamburger Socialverwaltung," Peter Jensen, 1990.

Fink, A. E. "Causes of Crime, Biological Theories in the United States, 1800-1915," New York: Barnes, 1938, 1962.

Fisher, R. A. [See biography by Box, Joan Fisher].

Fisher, R. A. "The Genetical Theory of Natural Selection," London: Oxford, 1930.

Fisher, R. A. "The Social Selection of Human Fertility," Oxford: Oxford [Clarendon Press], 1932.

Fisher, R. A. "Natural Selection, Heredity, and Eugenics, Including Selected Correspondence of R. A. Fisher with Leonard Darwin and Others," New York: Oxford, 1983.

Flournoy, T. "Spritism and Psychology," translated from the French by H. Carrington. New York: Harper, 1911.

Forrest, D. W. "Francis Galton: the Life and Work of a Victorian Genius," New York: Taplinger, 1974.

Foucault, M. and Binswager, L. "Dream and Existence," Special Issue of "Review of Existential Psychology and Psychiatry," Hoeller, K. (Ed.) 29, No 1, 1986.
Fowler, O. S. "Fowler's Practical Phrenology, etc.," New York: Fowler and Wells, The Phrenological Cabinet, 131 Nassau St., 1851.

Fowler, O. S. ""The Practical Phrenologist; and Recorder and Delineator of the Character and Talents," Boston,

Massachusetts, USA: O. S. Fowler, 514 Tremont St. 1869. Published at Riverside, Cambridge, Massachusetts, USA, by Houghton and Company.

Fowler, O. S. "Human Science or Phrenology; its Principles, Proofs, Faculties, Organs, Temperaments, Combinations, Conditions, Teachings, Philosophies, etc. etc. as Appled to Health, its Value, Laws, Functions, Organs, Means, Preservation, Restoration, etc. Mental Philosophy, Human and Self Improvement, Civilization, Home, Country, Commerce, Rights, Duties, Ethics, etc., God, his Existence, Attributes, Laws, Worship, Natural Theology, etc. Immorality, its Evidences, Conditions, Relations to Time, Rewards, Punishments, Sin, Faith, Prayer, etc., Intellect, Memory, Juvenile and Self-Education, Literature, Mental Discipline, The Senses, Sciences, Arts, Avocations, A Perfect Life. etc etc.," no place or publisher, 1874.

Flugel, J. C. "A Hundred Years of Psychology, 1833-1933," London: Gerald Duckworth, 1933.

Freeland, G. E. and Adams, J. T., "America's World Backgrounds," Sacramento, California: California State Department of Education, 1936.

Freud, S. "The Interpretation of Dreams," [1900] Volumes 4 and 5, The Standard Edition of the Complete Psychological Works of Sigmund Freud. Translated from the German by James Strachey and others. London: Hogarth, 1953-1974.

Freud S. "New Introductory Lectures on Psycho-Analysis," [1932] Lectures 29-35. Volume 22. The Standard Edition of the Complete Psychological Works of Sigmund Freud.

Translated from the German by James Strachey and others. London: Hogarth, 1953-1974.

Freud, S. "Totem und Tabu, Einige Uberemstimmungen um Seelenleben der Wilden und der Neurotiker," Vienna: Internationale Psychoanalytischer, 1913. See also Freud, S. "Totem and Taboo, Resemblances between the Psychic Lives of Savages and Neurotics," London: Kegan, Paul, Trench, Trubner, 1916. See also Freud, S. "Totem and Taboo," [1913] Volume 13. The Standard Edition of the Complete Psychological Works of Sigmund Freud". translated from the German by James Strachey and others. London: Hogarth, 1953-1974.

Fuller, R. C. "Mesmerism and the American Cure of Souls," Philadelphia, Pennsylvania, USA: University of Pennsylvania Press, 1982. Gall, F. J. "Anatomnie et Physologie du Systeme Nerveux au General, et du Cerceau en Particulier," Paris: Schoell, 1810.

Gall, F. J. "Sur les Fonctions," (Three Volumes]: no publisher or place, 1822.

Seemingly the original of "On the Functions of the Cerebellum. Gall, Vimont,
and Broussais" [1838] Translated from the French by George Combe. [See Combe]

Galton, F. "Inquiries into Human Faculty and its Development," London: Macmillan, 1883.

Galton, F. "Natural Inheritance," London: no publisher listed, 1889.

Galton, F. "Hereditary Genius," London: Macmillan, 1869 (Second Edition, 1892.

Galton, F. "The Average Contribution of Each Several Ancestor to the Total Heritage of the Offspring," Proceedings of the Royal Society, 1897, 67, p. 401.

Galton, F. "Essays in Eugenics," London: Eugenics Education Society, 1909. Gibbs-Smith, C. H. "The Great Exhibition of 1851," London: Victoria & Albert Museum, 1950, 1964.

Gillham, N. W. "A Life of Sir Francis Galton, from African exploration to the birth of eugenics," Oxford, UK: Oxford, 2001.

Glad, J. "Jewish Eugenics," Washington DC, Wooden Shore, 2011.

Glickman, S. E. "Some Thoughts on the Evolution of Comparative Psychology," in "A Century of Psychology as Science," Koch, S and Leary, D. E. [Eds.] Washington DC: American Psychological Association, 1992.

Glover, E., Mannheim, H. [Editor of the volume] and Miller, E., "Pioneers in Criminology," Number 1. Chicago, Illinois, USA: Quandrangle Books, 1960. Pages 168-227 contain information on Cesare Lombroso, this a chapter by Marvin E. Wolfgang.
Glueck, S. "Crime and Correction: Selected Papers," Cambridge, Mass.: Addison-Wesley, 1952.

Glueck, S. and Glueck, E. "Unraveling Juvenile Delinquency," New York: Commonwealth Fund, 1950.

Glueck, S. and Glueck, E. "Physique and Delinquency," New York: Harper, 1956.

Glueck, S. and Glueck, E."Predicting Delinquency and Crime," Cambridge, Massachusetts, USA: Harvard Press, 1959, 1967.

Goddard, H.H. "Heredity of Feeble-mindedness," American Breeders Magazine, Washington DC, USA: 1910, 1, Number 3. Reprinted in Eugenics Record Office, Bulletin Number 1: "Heredity of Feeble-mindedness," 1911.

Goddard, H. H. "The Kallikak Family: a Study in the Heredity of Feeble-mindedness," New York: Macmillan, 1912.

Goddard, H. H. "Sterilization and Segregation," New York: Russell Sage Foundation, 1913.

Godwin, G. S. "Criminal Man," New York: Braziller, 1957.

Goldsmith, M. "Franz Anton Mesmer, A History of Mermerism," Garden City, New York: Doubleday, Doran, 1934.

Goring, C. "The English Convict, a Statistical Study." His Majesty's Stationery Office. London, 1913. [n.s.]
Gosney, E. S. "Sterilization for Human Betterment; A Summary of Results of 6000 Operations in California, 1909-1929," New York: Macmillan, 1929. [Introduction by Popenoe, P. B.] who is in some references given credit as first author.] A collection of pamphets, some by Gosney, some by Popenoe, some by others.

Gould, S. J. "The Mismeasure of Man," New York: Norton, 1981. "Revised and Expanded Edition", 1996.

Grant, M. "The Passing of the Great Race, A Racial Basis of Euopean History," New York: Charles Scribner, 1916.

Grant, M. "Preserve an American Worth Fighting For," Berkeley, California: Save the Redwood League, 1921.

Grant, M. "The Alien in our Midst: or Selling our Birthright for a Mess of Pottage," New York: Galton Press, 1930.

Grant, M. [Introduction by Henry Fairfield Osborn]. "The Conquest of a Continent, or the Expansion of Races in America," New York: Charles Scribner's Sons, 1923. Republished, New York: Arno Press, New York Times, 1977.

Gray, R. T. "About Face, German Physiognomic Thought from Lavater to Auschwitz," Detroit, Michigan: Wayne State Press, 2004.

Gutting, G. "Michel Foucault's Archeology of Scientific Reason," Cambridge, UK: Cambridge Press, 1989.

Hall, C. S. "Psychology, an Introductory Textbook," Cleveland, Ohio, USA: Howard Allen, 1960.
Hall, G. S. "Founders of Modern Psychology," New York: Appleton, 1924. Hamilton, E. "The Greek Way," New York: Norton, 1930.

Hall, P. F. "Immigration and its Effects upon the United States," [Second revised Edition] New York: Holt, 1908.

Haller, M. H. "Eugenics, Hereditarian Attidudes in American Thought," New Brunswick, New Jersey, USA: Rutgers University Press, 1963.

Hamer, D. H. "The God Gene: How faith is hardwired into our genes," New York: Doubleday, 2004.

Hamer, D. H. and Copeland, P."The Science of Desire: The Search for the gay gene and the biology of behavior," New York: Simon & Schuster, 1994.

Harper, K. "Give me my Father's Body, the Life of Minik, the New York Eskimo", Iqaluit [Frobisher Bay, North West Territories (XOA OHO)] Canada: Blacklead Books, 1986.

Hartl, E. M., Monnelly, E. P., and Elderkin, R. D. "Physique and Delinquent Behavior, A Thirty-year Follow-up of William H. Sheldon's Varietes of Delinquent Youth," New York: Academic, 1982.

Hartmann, E. von. "Philosophie des Unbewussten," Volumes 1-3, Leipzig: Dietmar Klotz, 1923.

Hartzler, H. B. " (Rev.) Moody in Chicago or the World's Fair Gospel Campaign, Six Months' Evangelistic Work in the City of Chicago and Vicinity during the Time of the World's Columbian Exposition, Conducted by Dwight L. Moody and his Associates," New York: Fleming H. Revell, 1894.

Heizer, R. F. and Kroeber, T(heodora). (Eds.) " Ishi the Last Yahi, a Documentary History," Berkeley and Los Angeles: California Press, 1979.

Hemleben, J. "Rudolph Steiner, a Documentary Biography," East Grinstead, Sussex, UK: Henry Goulden, 1975. (Presumably 'Rudolph' is the British spelling of ther proper name 'Rudolf'.)

Henle, M., Jaynes, J. and Sullivan, J.J. (Eds.) "Historical Conceptions of Psychology," New York: Springer, 1973.

Herbert, W. "Across the Top of the World, the Last Great Journey on Earth," New York: Putnam's sons, 1971.

Herbert, W. "The Noose of Laurels, Robert E. Peary and the Race to the North Pole," New York: Atheneum, 1989.

Herman, E. "The Romance of American Psychology, Political Culture in the Age of Experts," Berkeley, California, USA: University of California Press, 1996.

Hodgson, G. M. and Knudsen, T. "Darwin's Conjecture, the search for general principles of social and economic evolution." Chicago: University of Chicago Press, 2010.

Hofstadter, D. R., "Metamagical Themas: Questing for the Essence of Mind and Pattern," New York: Basic Books, 1985.

Hofstadter, D. R. and Dennett, D. C., "The Mind's I," New York: Basic Books, 1981.
Holmes, S. J."A Bibliography of Eugenics," Berkeley, California: Press 1924. Holton, G. "Thematic Origins of Scientific Thought; Kepler to Einstein," (Revised Edition) Cambridge, Mass., USA: Harvard, 1988.

Hooton, E. A., "Crime and The Man," Cambridge, Massachusetts, USA: Harvard Press, 1931.

Hooton, E. A., "Up from the Apes," New York: Macmillan, 1931.

Hooton, E. A., "Apes, Men, and Morons," New York: Putnam, 1937.

Hooton, E. A., "Man's Poor Relations," Garden City, New York: Doubleday, 1942.

Hooton, E. A. With the collaboration of the statistical laboratory of the division of anthrophology, Harvard University. "The American Criminal, An Anthropological Study, Volume 1, The Native White Criminal of Native Parentage," Cambridge: Massachusetts, USA.: Harvard Press, 1939.

Horgan, J. "The End of Science," Boston, MA: Little Brown, 1997. Horgan, J. "Rational Mysticism," Boston, MA. Houghton Mifflin, 2003. Hrdlicka, A. "Physical Anthropology, its Scope and Aims, its History and Present Status in the United States," Philadelphia: Wistar Institute of Anatomy and Biology, 1919.

Humphrey, N. "Consciousness Regained," Oxford, UK: Oxford Press, 1983.
Ives, H. C. [Introduction by] "The Dream City, a Portfolio of the World's Columbian Exposition," St Louis, Missouri, USA: Thompson, 1893.

James, P. D. "The Murder Room," new York: Vintage (Random House), 2003.

James, W. "The Principles of Psychology," Two Volumes. New York: Holt, 1890.

Jefferis, B., G. and Nichols, J. L. "Light in Dark Corners: A Complete Sexual Science and Guide to Purity," New York: Grove, 1967. First published, 1887. Jung, C. "The Archetypes and the Collective Unconscious," Princeton, New Jersey, USA: Princeton, 1968.

Kagan, J. "Unstable Ideas: Temperament, Cognition, and Self," Cambridge, Massachusetts, USA: Harvard. 1989.

Kant, I. "Dreams of a Sprit Seer," Translated from the German by E. F. Goerwitz. London: Sonneschein. New York: Macmillan, 1900.

Kant, I. "Critique of Pure Reason," [translated from the German by Norman Kemp Smith] London: Macmillan; New York; St Martin's, 1961."

Kaufman, M. T. "A Museumís Eskimo Skeletons and its Own," Article from the New York Times, August 21, 1993.

Kaupen-Haas, H. (Ed.) "Der Griff nach der Bevolkerungspolitik, Aktualititat und Kontinuitit nazistischer Bevolerungspolitik. Nordlingen: Greno, 1986.

Kern, J. and Hammerstein, O. "Showboat," Book and Lyrics: © 1927, 1934, 1988.
Keynes, R. D. (Ed.) "Charles Darwin's Beagle Diary," Cambridge, UK: Cambridge Press, 1988.

Kevles, D. J. "In the Name of Eugenics," New York: Knopf, 1985.

Koch, L. "Racehygienje i Danmark 1920-1956," Copenhagen: Nordisk Forlag, 1996.

Kretschmer, E. "Physique and Character; An Investigation of the Nature of Constitution and of the Theory of Temperament," Second Edition: Revised by E. Miller. New York: Humanities Press, 1951. [Based on the 1925 edition.]

Kretschmer, E. "Korperbau und Charakter; Untersuchungen zum Konstitutions-Problem und zur Lehre von den Temperamenten," Berlin: J. Springer, 1922.

Kroeber, K. and Kroeber, C. (Eds.) "Ishi in Three Centuries," Lincoln, Nebraska, USA: Nebraska Press, 2003.

Kroeber, T(heodora). "Ishi, in Two Worlds: A Biography of the Last Wild Indian in North America," Berkeley and Los Angeles: University of California Press, 1961.

Kummer, H. "Social Organization of Hamadryas Baboons, a Field Study," Basel and New York: Karger, 1968.

Kummer, H. "In Quest of the Sacred Baboon, A Scientist's Journey." Translated from the German by M. Ann Biederman-Thorson. Princeton, New Jersey, USA: Princeton Press, 1995.

Kusch. M. "Foucault's Strata and Fields, An Investigation into Archeological and Genealogical Science Sudies," Dordrecht: Kluwer, 1991.

Lachman, Gary. Rudolph Steiner, an Introduction to his Life and Thought. New York: Tarcher (Penguin), 2007.

Lasswitz, K. G. "Th. Fechner, " Stuttgart: Frommanns, 1902.

Laughlin, H. H. "Eugenics Record Office. Report No. 1," Long Island, New York, USA: Cold Spring Harbor, June 1913.

Laughlin, H.H. "Report of the Committee to Study and to Report on the Best Practical Means of Cutting off the Defective Germ-Plasm in the American Population," Bulletin 10A, Eugenics Record Office, Cold Spring Harbor, Long Island, New York, USA, February 1914.

Laughlin, H. H. "Eugenical Sterilization in the United States," Chicago: Psychopathic Laboratory of the Municipal Court of Chicago, 1922.

Laughlin, H. H. "Eugenical Sterilization; 1926; Historical, Legal, and Statistical Review of Eugenical Sterilization in the United States," New Haven, Connecticut, USA: The American Eugenics Society, undated, circa 1926.

Le Goff, J. "Histoire et Memoire," Paris: Gallimard, 1988.

Le Goff, J. and Nora, P. (Eds.) "Constructing the Past, Essays in Historical Methodology," Cambridge, UK: Cambridge, 1985. Introduction by Colin Lucas. Translated from the French, "Faire de l'histoire, Nouveaux Objects," Chapter 8, "Mentalities: A History of Ambiguities" pp 166-180 is by J. Le Goff. Paris: Gallimard, 1974.

Lillard, A. S. "Other Folks: Theories of Mind and Behavior," Psychological Science, 1997, 8, 268-274.

Lindner, R. "Rebel without a Cause," New York: Grune and Stratton, 1944.

Lindner, R "Stone Walls and Men" New York: Odyssey, 1946.

Lippman, W. Tests of Hereditary Intelligence," "The New Republic", November 22, 1922, pp 318-330.

Lombroso, C. "L'uomo Delinquente," First Edition 1876 through the fifth edition, 1897. Turin, Italy: Bocca.

Lombroso, C. "Criminal Man, Considered in Relation to Anthropology, Jurisprudence, and Psychiatry," no publisher listed, 1911.

Lombroso, C. "Crime, its causes and remedies," Little-Brown, 1911. (trans. H. P. Horton) Reprinted by Patternson Smith, 1968.

Lombroso, C. "After Death — What? Spiritistic phenomena and their Interpretation," 'Rendered' into English by William Sloane Kennedy. Boston, Massachusetts, USA: Small, Maynard & Co.
Lombroso-Ferrero, G. "Criminal Man, According to the Classification of Cesare Lombroso, Briefly Summarised by his Daughter Gina Lombroso-Ferrero," New York: Putnam, 1911.

Lombroso, C. and Ferrero, W. "The Female Offender," Littleton, Colorado, USA: Rothman, 1980. [reprint].

Lowrie, W. [Edited and Translated from the German by W. Lowrie] "Religion of a Scientist," New York: Pantheon, 1946.

Ludwig, E. "Genius and Character," [translated from the German by Kenneth Burke]. New York: Blue Ribbon Books, 1927.

Lumsden, C. J., and Wilson, E. O., "Genes, Mind, and Culture: The Coevolutionary Process," Cambridge, Massachusetts, US: Harvard, 1981.

Lynch, A. "Thought Contagion, How Belief Spreads through Society, The New Science of Memes," New York: Basic Books, 1996.

Malina, J. B. See Mutter.

Macey, D. "The Lives of Michel Foucault, a Biography," New York: Pantheon, 1993.

Maier, B. N. "The Role of James McCosh in God's Exile from Psychology," History of Psychology, 2004, 7, 323-339.

Malone, J. C. "Psychology, Pythagoras to present. Cambridge, Massachusetts: MIT Press, 2009.
Marshall, M. E. and Wendt, R. A. "Wilhelm Wundt, Spritism, and the Assumptions of Science," (n.s.)

Mayer, J. "Gesetzliche Unfruchtbarmachung Geisteskranker," Freiburg: Herder, 1927.

Mesmer, F. A. "Prècis Historique des Faits Relatifs au Magnètisme Animal jusques en Avril 1781," London: 1781. Cited by Ettinger, 1970. No publisher's name given.

McCauley, R. N. (ed.] "The Churchlands and their Critics," Oxford, UK: Blackwell, 1996.

McCosh, J. "The Emotions," New York: Scribner's Sons, 1880.

McCosh, J. (see Sloan)

McDougal, W. "Is America Safe for Democracy?" New York: Arno, 1977.

McLaren, A. "Our own Master Race, Eugenics in Canada 1885-1945," Toronto: McCelland and Stewart, 1990.

Miller, J. W. "In Defense of the Psychological," New York: Norton, 1983. Mises, Dr. See under Fechner, G.T.

Mithen, S. "The Prehistory of the Mind, The Cognitive Origins of Art, Religion and Science," New York: Thames and Hudson, 1996.

Moody, D. (see Hartzler.)

Murchison, C. A. "Criminal Intelligence," Worcester, Mass.: Clark Press, 1926. Muccigrosso, R. "Celebrating the New World: Chicago's Columbian Exposition of 1893", Chicago, US, Ivan R. Dee, 1933.

—--Mutter, Ein Buch der Liebe und der Heimat fur alle," Photographs and quotations by J. B. Malina. Berlin: m. b. H., 1934.

—--"National Conference on Race Betterment," Second edition. Second National Conference on Race Betterment. San Francisco, California, August 4-8, 1915.

Nacquart, J. B. "Traite sur la Nouvelle Physiologie du Cerveay, ou, Exposition de la doctrine de Gall sur la

Structure de Beaucoup de Notes sur Differens Points de cette Doctrine, et Orne de Plances," Paris: Collin, 1808.

Nearing, S. "The Super Race: An American Problem," New York: Huebsch, 1912.

Nearing, S. "Social Sanity, A Preface to the Book of Social Progress," New York: Moffat, Yard, 1913.

Nbisonin, D. and Lindee, M. S. "The DNA Mystique, The Gene as a Cultural Icon," New York: Freeman, 1995.

—--New York Times, articles from, see Daley, see Kaufman.

Noble, W. and Davidson, I.,"Human Evolution, Language and Mind," Cambridge, UK: Cambridge Press, 1996.

Noll, R. "The Jung Cult, Origins of a Charismatic Movement," Princeton, New Jersey, USA: Princeton, 1994.

Oelze, B. "Gustav Theodor Fechner, Seele und Beseelung," Waxmann Munster: New York, 1989.

Osborn, H. F. "Are Acquired Variations Inherited? An Argument by Henry Fairfield Osborn Opening a Discussion upon the Lamarckian Principle," Philadelphia: no publisher, 1911.

Osborn, H. F. "The American Museum of Natural History, Its Origin, Its History, the Growth of its Departments to December 11, 1909," New York: Irving, 1911.

Osborn, H. F. "Men of the Old Stone Age, their Environment, Life and Art." New York, NY, USA:

Scribner's Sons, 1915. [The Hitchcock Lectures of the University of California, Berkeley, in 1914.]

Osborn, H. F. "The Hall of the Age of Man," New York: The American Museum of Natural History, Guide Number, 52, May 1925.

Osborn, H. F., "Fifty Years of Princeton '77, a Fifty-four Year record of the Class of 1877 of Princeton College," Princeton, New Jersey, USA: Princeton Press, 1927.

Osborn, H. F. "Fifty-two Years of Research, Observation and Publication, 1877-1929; A Life Adventure in Breadth and Dept,." New York: Scribner's Sons, 1930.

Osborn, H. F. "Cope: Master Naturalist; the Life and Letters of Edward Drinker Cope," Princeton: Princeton Press, 1931.

Osborn, F. "Preface to Eugenics," New York: Harper, 1940.
Pepper, S. C. "World Hypotheses, A Study in Evidence," Berkeley and Los Angeles, California: California Press, 1942.

Pearson, K. "On the Inheritance of the Mental and Moral Characters in Man, and its Comparison with the Inheritance of the Physical Characteristers," Journal of the Anthropological Institute of Great Britain, 1903, 33, p. 179.

Pearson, K. "The Groundwork of Eugenics," London: Dular and Company, 1909.

Pearson, K. "Darwinism, Medical Progress and Eugenics," The Cavendish Lecture. London: Dulau and Company, 1912.

Pearson, K. "Francis Galton, 1822-1922, A Centenary Appreciation," London: Cambridge Press, undated.

Pearson, K., "Charles Darwin, 1809-1882, Being a Lecture Delivered to the Teachers of the London County Council, March 21, 1923," London: Cambridge Press, Undated.

Pearson, K. "The Right of the Unborn Child," Lecture of November 13, 1926. London: Cambridge Press, 1927.

Pearson, K. "The Life, Letters, and Labours of Francis Galton," Cambridge, UK: Cambridge, 1914-1930.

—-- [Phrenological Society] Transactions of the Phrenological Society, Instituted 22d February 1820, Edinburgh: Anderson, 55, North Bridge Street, 1824.
Pierce, J. W. "Photographic History of the World's Fair and Sketch of the City of Chicago. A Guide to the World's Fair and Chicago," Baltimore, Maryland, USA: Woodward, 1893.

Pinker, S., "How the Mind Works," New York: Norton, 1997.

Pliny [Caius Plinius]. "Natural History," Volume 3: Cambridge, Massachusetts, USA: 1940.

Pommwein, R. "Sterilisierung der Rheinlandbastarde, Das Schicksal einer farbigen deutschen Minderheit 1918-1837," Dusseldorf: Droste, 1979.

Popenoe, P. B. "The Conservation of the Family," Baltimore, Maryland, USA: The Williams and Wilkins Company, 1926.

Popenoe, P. B. "Collected Papers on Eugenic Sterilization in California; a Critical Study of Restults in 6000 Cases," Pasadena, California. USA: The Human Betterment Foundation, 1930. [See Gosney, E. S.]

Popenoe, P. B. and Gosney, E. S., "Twenty-eight years of Sterilization in California," Pasadena, California; USA: The Human Betterment Foundation, 1938.

Popenoe, P. B. and Johnson, R. H. "Applied Eugenics," New York: Macmillan, 1933.

Puységur, J. M. P. de C., Compte de. "Suite des Memoires pour Aervir l'histoire et l'Èstablissement du Magntisme Animal." Cited by Crabtree, 1993, p.397.

Quammen, D. "The Song of the Dodo: Island Biogeography in an Age of Extinctions," New York: Scribner, 1996.

Rainger, R. "An Agenda for Antiquity, Henry Fairfield Osborn and Vertebrate Paleontology at the American Museum of Natural History, 1890-1935", Tuscaloosa, Alabama, USA: University of Alabama Press, 1991.

Raepke, C. O. "The Evolution of Progress," New York: Random House, 1993.

Rasku, I. and Downes, C. S. "Genes in Medicine, Molecular Biology and Human Genetic Disorders," New York: Chapman and Hall, 1995.

Redfield, C. L. "Great Men and How They are Produced," published by the author, Chicago, 1915.

Reich, W. "Cosmic Superimposition, Man's Orgonotic Roots in Nature," Orgonon, Rangley, Maine: The Wilhelm Reich Foundation, 1951.

Reich, W. "Selected Writings, An Introduction to Orgonomy," New York: Farrar, Straus, and Giroux, 1973.

Rembis, M. A. "I Ain't Been reading While on Parole": Experts, Mental tests, and Eugenic Commitment Law in Illinois, 1890-1940. History of Psychology, 2004, 7, 225-247.

Richter, V. "Literature after Darwin, Human beasts I n western fiction, 1859-1939." New York: Palgrave Mcmillan, 2011.

Ritvo, L. B. "Darwin's Influence on Freud, a Tale of Two Sciences," New Haven, Connecticut, USA: Yale Press, 1990.

Rosenbaum, R. "The Great Ivy League Nude Picture Photo Scandal: How Scientists Coaxed America's Best and Brightest out of their Clothes," New York: The New York Times Sunday Magazine, January 15, 1995.

Ruse, M. "The Darwinian Revolution," Chicago: Chicago Press, 1979.

Rydell, R.W. "All the World's a Fair, Visions of Empire of American International Expositions, 1876-1916," Chicago, Illinois. USA: Chicago, 1984.

Rydell, R. W. "World of Fairs, The Century -of-Progress Expositions," Chicago, Illinois, USA: Chicago, 1992.

Sahakian, W. S. (Ed.) "History of Psychology, A Source Book in Systematic Psychology," Itasca, Illinois, USA: Peacock, 1969.

Schiller, F. C. S."Eugenics and Politics, Essays by Ferdinand Canning Scott Schiller," Boston and New York: Houghton Mifflin, 1926.

Searle, G. R. "Eugenics and Politics in Britain, 1900-1914," Leyden: The Netherlands: Noordhoff, 1976.

Shaw, M. "World's Fair Notes, A Woman Journalist Views Chicago's 1893 Columbian Exposition," no city given: Pogo Press, 1992.

Sheldon, W. H. "Psychology and the Promethean Will," New York. NY, USA: Harper, 1936.

Sheldon, W. H. "(with the collaboration of S. S. Stevens and W. B.Tucker). The Varieties of Human Physique: An Introduction to Constitutional Psychology.
New York: Harper, 1940.

Sheldon, W. H. "Prometheus Revisited," Cambridge, Massachusetts, USA Schenkman, 1974.

Sheldon, W. H. (with the collaboration of S. S. Stevens) "The Varieties of Temperament: A Psychology of Constitutional Differences," New York: Harper,
1944.

Sheldon, W. H. [with the collaboration of E. M. Hartl and E. McDermott] "Varieties of Delinquent Youth: An Introduction to Constitutional Psychiatry," New York: Harper, 1949. [See also Hartl].

Sheldon, W. H. [with the collaboration of C. W. Dupertuis and E. McDermott] "Atlas of Men: A Guide for Somatotyping the Adult Male at All Ages," New York: Harper, 1954.

Schama, S. "Landscape and Memory," New York: Knopf/Random, 1995.

Simms, M, "Darwin's Orchestra, An Almanac of Nature in History and the Sciences," 1997.

Simms, J. "Physiognomy Illustrated or, Nature's Revelations of Character," New York: Murray Hill, 129 E 28th St., New York City, NY, USA, 1887.
Sinnett, A. P. "The Rationale of Mesmerism," Boston and New York: Houghton, Mifflin, 1892.

Sizer, N. and Drayton, H. S. "Heads and Faces and How to Study Them; A Manual of Phrenology and Physiogomy for the People," New York: Fowler and Wells, 1889.

Sloan, W. M. [Ed.] "The Life of James McCosh, A Record Chiefly Autobiographical," Bristol, UK: Thoemmes, 1992. First published 1896, New York: Scribner's Sons.

Smith, J. D.and Nelson, K. R., "The Sterilization of Carrie Buck," Far Hills: New Jersey, USA: New Horizon Press, 1989.

Smith, J. D. "Minds Made Feeble: the Myth and Legacy of the Kallikaks," Rockville, Maryland, USA: Aspen Systems, 1985.

Sokal, M. S. (Ed.) "Psychological Testing and American Society, 1890-1930," New Brunswick, New Jersey: Rutgers, 1987.

——"Spiritualism," [Ed: Ward, G. L.] Three Volumes. New York: Garland, 1990. Sperber, D."Explaining Culture, a Naturalistic Approach," Oxford, UK:
Blackwell, 1996.

Sperber, D., Premack, D., and Premack, A. J., "Casual Cognition, a Multidisciplinary Debate," Oxford, UK: Clarendon, 1995.

Spurzheim, J. G. "Phrenology, in Connexion with the Study of Physiognomy," Boston, Massachusetts, USA: Marsh, Capen, and Lyon, 1826, 1833.

Steiner, R. "Die Philosophie der Freiheit," Berlin: Philosophisch Anthroposophischen, 1918. First published 1894. Published in English, beginning with the 1922 edition, as "The Philosophy of Freedom." Rudolph (sic) Steiner Press, 35 Park Road, London.

Steiner, R. "The Essentials of Education," London: Anthroposophical Publishing Company, 1926.

Steiner, R. "Knowledge of the Higher Worlds, How is it Achieved," (Translated from the German by D.S.O. and C.D.) London: Rudolph Steiner Press, 1969.

Steiner, R., "Grundlinien einer Erkenntnistheorie der Goetheschen Waltanschauung," [Translated by O. D. Wannamaker] New York:
Anthroposophic Press, 1978.

Steiner, R. and Wegman, I. "Fundamentals of Therapy, An Extension of the Art of Healing through Spritual Knowledge," [Translated from the German by George Adams.] Letchworth, Hertfordshire, UK: The Rudolph Steiner Press, The Garden City Press, 1925; 1967 (Revised).

Stevens, S. S., "The Surprising Simplicity of Sensory Metrics," American Psychologist, 1962, 17, 29-39.

Stich, S. P., [Ed,] "Innate Ideas," Berkeley, California, USA: University of California Press, 1975.
Stich, S. P. "From Folk Psychology to Cognitive Science, The Case against Belief," Cambridge, Massachusetts, USA: Massachusetts Institute of Technology Press, 1983.

Stich, S. P., "Deconstructing the Mind," New York: Oxford Press, 1996. Stich, S. and Nichols, S. "Folk Psychology: Simulation or Tacit Theory?" Mind and Language, 1992, 7, 35-71.

Strasburger, E. "Die Stofflichen Grundlagen der Organischen Verebung," Jena: no publisher, 1905.

Strelau, J. and Angleitner, A. (Eds). "Explorations in Temperament; International Perspectives on Theory and Measurement," New York: Plenum, 1991.

Subotnok, R. F. and Arnold, K. D. (Eds.) "Beyond Terman: Contemporary Longitudinal Studies of Giftedness and Talent," Norwood, New Jersey, USA: Ablex, 1994.

Swedenborg, E. "The Universal Human and Soul-Body Interaction," Edited and Translated by G. F. Dole. New York: Paulist Press, 1984.

Taylor, A. "Annie Besant, A Biography," Oxford and New York: Oxford Press, 1992.

Terman, L. M. "Mental and Physical Traits of a Thousand Gifted Children," Volume 1 in Terman, L. M. "Genetic Studies of Genius."

—Terman, L. M. volume ii. "Genetic Studies of Genius," See Cox, C. M.

— Terman, L. M. volume iii in "Genetic Studies of Genius," See Burks, et al. Terman, L. M. and Oden, M. H. " Gifted Child Grows up, Twenty-five years' (sic) Follow-up of a Superior Group," volume iv of "Genetic Studies of Genius," Stanford, California: Stanford, 1947.

Terman, L. M. and Oden, M. H. "he Gifted Child at Mid-life: Thirty-five Years Follow-up of the Superior Child," volume v of " Genetic Studies of Genius," Stanford, California: Stanford Press, 1925.

Terman, L. M. "The Gifted Child Grows Up, Twenty-five Years' Follow-up of a Superior Group," volume iv in "Genetic Studies of Genius." Stanford, California, USA: Stanford Press, 1947.

Torrey, E. F. "Out of the Shadows, Confronting America's Mental Illness Crisis," New York: John Wiley, 1997.

Trivers, R. "Social Evolution," Menlo Park, CA. Benjamin/Cummings, 1985.

Trobridge, G . "Life of Emanuel Swedenborg," New York: New Church Press,
1928.

Tulchin, S. H. "Intelligence and Crime, A Study of Penitentiary and Reformatory Offenders," Chicago, Illinois, USA: Chicago Press, 1939.

Vold, G. B. "Theoretical Criminology," New York: Oxford Press, 1958.

Wallace, A. R. "The Wonderful Century, Its Successes and Failures," New York: Dodd, Mead, 1899.
Warren, M. A."Gendercide: The Implications of Sex Selection," Totawa, New Jersey, USA: Rowman and Allanheld, 1985.

Washington, P. "Madame Blavatsky's Baboon, A History of the Mystics, Mediums, and Misfits who Brought Spritualism to America," New York: Schoken, 1993.

Wehr, G. "Rudolph Steiner, Leben-Erkenntnis Kulturimpuls," Munchen, Germany: Kesel, 1987.

Wells, S.R. "New Physiogomy or, Signs of Character, as Manifested through Temperament and External Forms, and Especially in "The Human Face Divine," New York: Fowler and Wells, 1867.

Westoby, A. "The Ecology of Intentions: How to make Memes and Influence People: Culturology," Foreword by D. C. Dennett, Center for Cognitive Studies, Tufts University. Draft distributed by internet.cogsci.soton.ac.uk/py104/dennett.rob.html

Wilson, E. O. "Sociobiology," Cambridge, Massachusetts, USA: Harvard Press 1975.

Wilson, E. O. "Consilience, the Unity of Knowledge," New York: Knopf, 1998.

Wilson, R. A. "Boundaries of the Mind, the Individual in the Fragile Sciences. Cambridge, UK: Cambridge, 2004.

Wood, J. G. "Illustrated Natural History," Philadelphia: Crawford, 51 N. Ninth, 1883.

—-- [World's Columbian Exposition] "World's Columbian Exposition 1892-1893, Report of the President."

—-- [World's Columbian Exposition] Official Guide to the World's Columbian Exposition in the City of Chicago, State of Illinois, May 1 to October 16, 1893. Compiled by John J. Flinn. Chicago: The Columbian Guide Company, 358 Dearborn Street.

—-- [World's Columbian Exposition] "Pamphlets on Oakland" — Papers numbers 4 [Alameda County World Fair Association] 1893; and 5 [Publicity and Promotion] 1893. University of California, Bancroft Library, xF869.02.P2 V3:5.

Wundt, W. M. "Der Spiritismus: eine Sogenannte Wissenshaftliche Frage," Leipzig: Englemann, 1879.

Wundt, W. M. "Hypnotismus und Suggestion," Leipzig: Engelmann, 1892. Also "Hypnotisme et Suggestion; Etude Critique," Translated from the German by A. Keller, Fourth Edition] Paris: Alan, 1909.

Wundt, W. M. "Völkerpsychologie: Eine Untersuchung der Entwicklungsgesetze von Sprache, Mythus, und Sitte," various editions in varying numbers of volumes. See, e.g., Leipzig: Krener, 1914-1922.

Wundt, W. M. "Elements of Folk Psychology; Outlines of a Psychological History of the Development of Mankind," Translation from the German by Edward LeRoy Schaub. London: Allen & Unwin; New York: Macmillan, 1916. Wynne-Edwards, V. C. "Evolution through Group Selection," Oxford and Boston: Blackwell Scientific, 1986.

Yerkes, R. M. and LaRue, D. W. "Outline of the Study of the Self," Cambridge: Massachusetts, USA: Harvard Press, 1914.

Yochelson, S. and Samenow, S, E., "The Criminal Personality: Volume 1: A Profile for Change," New York: Jason Aronson, 1976.

Zusne, L. "Biographical Dictionary of Psychology," Westport Conn.: Greenwood Press, 1984.

END NOTES

References to books include the author, title, year of publication and page number when relevant. The full references are to be found in the Bibliography.

Chapter 1: The Archeopsychology of Mental Fossils

Prefatory quote is from Darwin, C. "On the Origin of Species," 1859. [Later editions dropped the 'On'.]

1. Le Goff, J. & Nora, P. "Constucting the Past, Essays in Historical Methodology.1985, p. 166, translated from the French.
2. LeGoff, J. "Histoire et Memoire", 1988, p. 166.
3. Le Goff, J. & Nora, P. "Constructing the Past", 1974, p. 169.
4. I have used three sources: Foucault, M. & Binswager, L. "Dreams and Existence", translated by Williams & Needleman, 1986; Gutting, G. "Michel Foucault's Archeology of Scientific Reason", 1989; and Kusch, M. "Foucault's Strata and Fields", 1991. I thank Professor Kerry Walters of Gettysburg College for his comments on Foucault's archeology.
5. Freud, S. "Totem und Tabu, Einige Uberemstimmungen um Seelenleben der Wilden und der Neurotiker," (Totem and Taboo) 1913.
6. Jung, C. "The Archetypes and the Collective Unconscious," Princeton, New Jersey, USA: Princeton, 1968.

7. Wundt, W. M. "Völkerpsychologie: Eine Untersuchung der Entwicklungsgesetze von Sprache, Mythus, und Sitte."

8. Schaub, E. translation of Wundt, W. "Elements of Folk Psychology." 1916.

9. Hall, C. "Psychology: an Introductory Textbook". 1960

10. Stitch, S. "Innate Ideas". 1975

11. Dawkins, R. "The Selfish Gene." [1976; 1989, second, new edition, this being the first edition with end-notes by which the author refreshes the reader on the ideas, themes, and memes expressed earlier in the first edition.]. See also Dawkins, M. S., Unraveling Animal Behaviour, second edition. London: Longmans, 1995. [In the US, John Wiley & Sons.]

12. Additionally, Wallace thought militarism and colonialism to be curses, seeing no differences between the Spanish plundering of the New World and its people and the British actions in Africa. He was appalled that poverty in Britain led to the death of children whose only crime had been to have been born to a poor family. He criticized the economic disparity resulting from capitalistic excess, as some had oil lamps, while others died in the cold for want of coal. He found the economic enslavement of women and children to be immoral. He also warned of the overuse of non-renewable resources: "The struggle for wealth, and its deplorable result have been accomplished by the reckless destruction of the stored-up projects of nature, which is even more deplorable because it is irretrievable. Not only have forest growths of many hundreds of years been cleared away, but the whole of mineral resources, the slow products of long-past eons of time and geological change, are being exhausted."

Chapter 2: Some Initial Excavations

1. Goldsmith, M. "Franz Anton Mesmer, A History of Mesmerism," 1934, p. 64.
2. Goldsmith, M. "Franz Anton Mesmer, A History of Mesmerism," 1934, p. 63.
3. Ellenberger, H. "The Discovery of the Unconscious," 1970, p. 71.
4. Crabtree, A, "From Mesmer to Freud, 1993," p. 28-29.
5. Fowler, O. S. "Practical Phrenology, a Reading of Mr. Phillips", 1869.
6. Fowler, O. S. "Practical Phrenology," p. 2-4.
7. Sizer, N. and Drayton, H. S. "Head and Faces," 1889, pp 19-20, 185-186.
8. McCosh's books are on metaphysics and emotion. His ideas deserve more explicit attention than he now receives. For example, he published of the theory of emotion now called the [William] James-Lange theory in 1880. See Sloan, 1896; reprinted 1992.
9. Sims, M, "Darwin's Orchestra, An Almanac of Nature in History and the Arts," 1997, pp. 260-261.
10. Sloan, W.M. (Ed.) "The Life of James McCosh," 1896.
11. Kretschmer, E. "Physique and Character," 1925, 1951, p. 3.
12. Sheldon, W. H., (with the collaboration of S. S. Stevens and W. B.Tucker). "The Varieties of Human Physique: An Introduction to Constitutional Psychology," New York: Harper, 1940. p. 26.
13. Sheldon, W.H., (with the collaboration of S. S. Stevens and W. B.Tucker). "The Varieties of Human Physique: An Introduction to Constitutional Psychology.
14. Sheldon, W.H., (with the collaboration of S. S. Stevens and W. B.Tucker). "The Varieties of Human Physique: An Introduction to Constitutional Psychology.

15-17. Lombroso, C.. "Criminal Man, Considered in Relation to Anthropology, Jurisprudence, and Psychiatry," no publisher listed, 1911.
18-19. Haller, M. H. "Eugenics, Hereditarian Attitudes in Amiercan Thought," p. 139.

Chapter 3. The Congress of Ideas

1. Gibbs-Smith, C. H. "The Great Exhibition of 1851," 1950, 1964.
2. World's Columbian Exposition, Official Guide, 1893.
3. World's Columbian Exposition; Official Guide, 1893.
4. Ambler, A. T. and Banta, M., The Invention of Photography and its Impact on Learning, 1899. Badger, R., "The Great American Fair," 1979.
5. World's Columbian Exposition, Official Guide, 1893.
6. For sources on the World's Columbian Exposition, I have used Badger (1979), Bolotin and Laing, 1992, Brown (1994), Burg, 1975, and Ives (1893) and the Official Guides listed under World's Columbian Exposition and Report of the President. The description of Moody's efforts regarding evangelicism is from Hertzler (1893). Information on the anthropological displays is found in the Official Guide.
7. Bridges, W. "Gathering of Animals, An Unconventional History of the New York Zoological Society," New York: Harper and Row, 1974.
8. Herbert, W. "Across the Top of the World," 1971, but see especially Herbert, W., "The Noose of Laurels, Robert E. Peary and the Race to the North Pole,". For a full account of Minik from the Eskimo viewpoint, see Harper, K. "Give me my Father's Body, the Life of Minik, the New York Eskimo." 1986, 2000. Only at the end of the book do we discover the relationship

between Minik and the author of this fine book. See also Kaufman, 1993.

9. Heizer, R. F. and Kroeber, T(heodora). (Eds.) "Ishi, the Last Yahi a Documentary History," 1979.p. 107. Kroeber, T[heodora]. "Ishi in Two Worlds," 1961. Heizer, R. F. and Kroeber, T., "Ishi in Two Worlds," 1979, p. 159.

10-13. Kroeber, T[heodora]. "Ishi in Two Worlds," 1961. Heizer, R. F. and Kroeber,

T., 1979, p. 242.

14. As the exact nature of James's exhibit seems not to be known, I have here used the collection of George Stratton, displayed at the psychology department of the University of California at Berkeley. Display: University of California at Berkeley, Tolman Hall.

15. See Rosenbaum, R., "The Great Ivy League Nude Posture Photo Scandal: How Scientists Coaxed America's Best and brightest out of their Clothes," New York Times Sunday Magazine, January 15, 1995. The exhibitionistic title obscures the point that the article takes up several issues of value; namely, whether such orders would be followed today and what protections there might be against the human misuse of other human beings. Such issues come at the end of an article otherwise devoted to the author's search for the original photographs. While decrying their availability, the New York Times article nonetheless features them as its cover on this issue and reproduces several in the article.

15-18. World's Columbian Exposition, Official Guide, 1893.

19. Sokal, M. S. (Ed.) "Psychological Testing and American Society", 1890-1930. Flugel, J. C." A Hundred Years of Psychology." 1833, 1933. World's Columbian Exposition, Official Guide, 1893.

20-22. Appelbaum, S. "A Photographic Record: Photos from the Collections of the Avery Library of Columbia University and the Chicago Historical Society," New York: Dover, 1980. H. C. Ives, "The Dream City, a Portfolio of the World's Columbian Exposition, 1893. Brown, J. K. "Contesting Images, Photography and the World's Columbian Exposition," 1994. Badger, R. "The Great American Fair," 1979, p. 107, 108.

23. Darwin, C. "The Expression of the Emotions in Man and Animals." 1872.

24-26. Forrest, D. W. Francis Galton: "The Life and Work of a Victorian Genius," Gillham, N. W. "A Life of Sir Francis Galton," 2001. Information regarding the South Kensington Exhibition is from Forrest, 1974, pp 181-186. Flugel, J. C. "A Hundred Years of Psychology, 1833, 1933. When the exhibit was closed at South Kensington, the equipment was given to Thompson at Oxford toward the establishment of an anthropological laboratory there. (Forrest, 1974, p. 182.) .

Galton's active imagination everywhere touches us. After a trip to (then) South West Africa he noted that oxen, although seeming to be individually competitive, nonetheless acted as a unit when separated. Thinking about the reason that some animals 'herd', he reached the conclusion that natural selection may favor 'group behavior'. This notion is superficially contrary to Darwin's insistence on selection at the level of the individual, but it is an idea that surfaces from time to time, most recently in Wilson's (1976) version of sociobiology (see Chapters 2-4], as well as in Trivers [1985] and Wynne-Edwards [1986].

Chapter 4. Spiritism and Unseen Worlds

General comment: The word 'spiritism', according to the Encyclopedia of Occultism and Parapsychology, has a different meaning from 'spiritualism.' The difference is not a firm one, but 'spiritism' refers to the belief that the spirits [or souls] of the dead can communicate to the living while 'spiritualism' usually embodies Christian theology. As 'spiritism' is the word most often used by the writers here to be considered, I have used it as well, although it is evident that Christian and Hindu teachings overlay their work. I am about to present the case that the our study of the human mind has a long history of using spiritist ideas as explanations, and that, perhaps surprisingly, experimental psychology has such origins.

1. Any discussion of the unconscious, writing as I am after the turn of the millenium, must fall not into the shadow, but into the brilliance of Henri F. Ellenbergerís "The Discovery of the Unconscious, the History and Evolution of Dynamic Psychiatry,"published in 1970. This monumental work so convincingly sets the case for the notion that the 'unconscious', as a concept, is as old as human thought but truly begins to effect psychiatry around 1775. The book of 11 chapters is, alternatively, eleven self-contained chapters, these on Janet, Freud, Adler, and Jung, along with several on Mesmer, the Puysegur brothers, and others. The intellectual history here contained is so stunning in depth and breadth that it is unlikely that anyone can do more, in our times, than add footnotes. My own footnote is relating dynamic psychiatry to spiritism and, perhaps, to the nineteenth century ideas of Fechner, Steiner, and Neitzsche. Ellenberger, H. F. "The Discovery of the

Unconscious, the History and Evolution of Dynamic Psychiatry," 1970, p.166.

2. Ellenberger, H. F. "The Discovery of the Unconscious, the History and Evolution of Dynamic Psychiatry," 1970, p. 171.

3. Ellenberger, H. F. "The Discovery of the Unconscious, the History and Evolution of Dynamic Psychiatry," 1970, p. 171.

4. Ellenberger, H. F. "The Discovery of the Unconscious, the History and Evolution of Dynamic Psychiatry," 1970, p. 171.

5. Let it be confessed that it is difficult, maybe impossible, to write objectively about Rudolph Steiner. First, my natural and cultural way of thinking about things is through the materialist filter, and I think one would have to have been emersed in the system to grasp, or intuit, the relationships of ideas he suggests. Second, Steiner was both philosopher and promoter. The library I use owns 245books and documents by him. Some are on-the-spot transcriptions by audience members of lectures he gave [this in the days of shorthand]; a few are books meant to explain and attract; others are charges to the already faithful. Repitition with so much variation makes it impossible for me to be certain that a view is fixed, thereby correctly explained.

The word, 'anthrosophistry' and its variants, refers to the 'wisdom of humankind'. The biblical source is Corinthians 2, especially 14-15.

Steiner, by the way, was no stranger to experimental psychology: he received a PhD approved by von Hartman, the philosopher, not the psychologist. Steiner was a direct, second-generation, descendent of the beginnings of experimental psychology by Fechner. Steiner, R. "Knowledge of the Higher Worlds, How is it

Achieved," (Translated from the German by "D.S.O. and C.D", 1969.

6. Ahern, G "Sun at Midnight," 1984. Ahern's book is unlike the usual hagiopgraphic account of Steiner. It is appreciative to be sure, and places Steiner within the history of esoteric thought. It includes an account of Anthroposophy as practiced by a small group in England in the 1970s, these interviews thereby giving anthroposophy a somewhat fuller and enriched meaning and coherent philosophy. Steiner, R. "The Essentials of Education," 1926. See also the more balanced explication by Lachman, G. "Rudolph Steiner", 2007.

7, Steiner, R. "The Essentials of Education," 1926.

8. Steiner, R. and Wegman, I. "Fundamentals of Therapy," 1925, p 13-14.

9. Steiner, R. and Wegman, I. "Fundamentals of Therapy," 1925, p. 17.

10.Steiner, R. and Wegman, I. "Fundamentals of Therapy,"1925, p. 57, 120-121.

11-13. Steiner, R. and Wegman, I. "Fundamentals of Therapy,"1925.

14. Zusne, L. "Biographical Dictionary of Psychology," 1984. Hall, G. S. "Founders of Modern Psychology," 1924.

15. Fechner, G. T. "Elemente der Psychophysik", 1860. Fechner, G. T. "Nanna, Oder uber das Seelenleben der Planzen," 1921. (Ed. K. Lasswitz), Fechner, G. T. "Uber die Seelenfrage. Ein Gangdurch die Sichtbare Welt, um die Unsichtbare zu Finden," 1907, Fechner, G. T. "Zend Avesta: oder, uber die Dinge de Himmels und des Jenseits vom Standpunkt der Naturbetrachting,"1922, Fechner, G. T. "Little Book of Life after Death," See Fechner, G. T., written as Dr. Mises, 1875. In addition to Fechner's works, most of

which are available both in German and English, I have consulted two biographies, these by Lasswitz and by Oelze. Two other works furnish powerful views of his works. G. Stanley Hall, the first American [by his own count] to study with Wundt, and therefore someone who knew the nineteenth century bevy of philosophers/physicists/physiologists/ psychophysicists [Zeller, Lotze, Fechner, Eduard von Hartmann [to whom Steiner's PhD thesis was dedicated], Helmholtz [1821-1894] and Wundt [1832-1920], has left us an account through his 1912 lectures at Columbia University on these 'Greats'. (Hall, 1924). Hall knew Fechner hardly at all; his memory of their meeting is blurry, but Hall's account of the development of Fechner's thought is handsomely done.

Walter Lowrie, a lifetime student of languages and religion, provides us with his translations from some of Fechner's works and with a running commentary on how they relate to Fechner's religion. The importance of this book is that we have few accounts by religionists as to how Fechner's work suits their interests, as psychology has appropriated Fechner's experimental side.

Lowrie unnerves this reader by a long opening chapter in which he explains that a publisher could not be found easily, as, in this chapter, Lowrie is at pain to point out that at last Darwinism is dead; that no decent college would hire a man who taught Darwinism [Lowrie spent his life in the shadows of the spires of Princeton University, where he had been a student under President McCosh, described in Chapter 2, who, it will be recalled, although a clergyman, brought the study of evolution to Princeton, if not to Lowrie — and taught the subject himself in the course in physiological psychology.] Nonetheless, Lowrie's views of Fechner is alone, or almost so, in being a view of psychophysics from the view of the 'other world' and it is therefore of compelling interest. Lowrie's home was to become the home of Princeton presidents.

15. Zusna, L. "Biographical Dictionary of Psychology", 1984.
16. Fechner, G. T. Elemente der Psychophysik, 1860.
17. Fechner, G. T. "Zend Avesta: oder, uber die Dinge de Himmels und des Jenseits vom Standpunkt der Naturbetrachting,"1922, Fechner, G. T. "Little Book of Life after Death,"
18. Freud, S. "Totem und Tabu,"Einige Uberemstimmungen um Seelenleben der Wilden und der Neurotiker," 1913. See also Freud, S. "Totem and Taboo, Resemblances between the Psychic Lives of Savages and Neurotics", 1916. See also Freud, S. "Totem and Taboo," [1913] Volume 13. The Standard Edition of the Complete Psychological Works of Sigmund Freud," 1953-1974.
19. Bakan, D. "Sigmund Freud and the Jewish Mystical Tradition," 1958, 1959.
20. Freud, S. The Standard Edition of the Complete Psychological Works of Sigmund Freud". translated from the German by James Strachey and others. London: Hogarth, 1953-1974.
21. Kummer, H., "Social Organization of Hamadryas Baboons," 1968.
22. Darwin, C. "The Expression of the Emotions in Man and Animals," 1872, 1965, p. 108.
23. Darwin, C. "The Expression of the Emotions in Man and Animals," 1872, 1965, p. 190.

Chapter 5: The Mental Fossil of Physiognomy; Character and Talent

1. Gillham, N. W. "A Life of Sir Francis Galton, from African exploration to the birth of eugenics," Oxford, UK: Oxford, 2001. Also from the exhibit "Police photoghraphy," Mounted by the Museum of Modern Art of San Francisco, seen at the Grey Gallery, New York University.
2. Fowler, O. S. and Wells, S. R., The Phrenological Cabinet," 1889. See, also Wells, S. R. "New Physiognomy, 1867; Sizer, N. and Drayton, H. S. "Heads and Faces," 1889; Davies, J. D., "Phrenology, Fads and Science." 1955. Gall, F. J. "Sur les Fonctions," (Three Volumes]: no publisher or place, 1822. Seemingly the original of "On the Functions of the Cerebellum. Gall, Vimont, and Broussais," [1838] Translated from the French by George Combe. [See Combe].
3. Wells, S. R. "New Physiognomy", 1867.
4. Acknernecht, E. H. and Vallois, H. V. "Franz Joseph Gall," 1956, p. 7. Transactions of Phrenology," — [Phrenological Society], 1820, p.679. Gall, F.J. Transactions of Phrenology," — [Phrenological Society] Transactionsof the Phrenological Society, Instituted 22d February 1820. Sizer, N. and Drayton, H. S. "Head and Faces," 1889, pp 19-20, 185-186.
5. Simms, J. "Physiognomy Illustrated or, Nature's Revelations of Character," 1887.
6. Wells, S. R. "New Physiognomy", 1867. The quotations regarding noses are all from Wells, S. R. "New Physiognomy," 1867, pp 215-218.
7. Simms, J. "Physiognomy Illustrated or, Nature's Revelations of Character," 1887.
8. Combe, G. "The Constitution of Man, Considered in

Relation to External Objects," 1849. (The original Edinburgh, Scotland edition is 1825.). Wallace, A. R. "This Wonderful Century,"1899. p 191-192.

Chapter 6: Physique, Temperament, and Madness

1. Ellenberger, H. F., "The Discovery of the Unconscious: The History and Evolution of Dynamic Psychiatry," 1970.
2. Kretschmer, E. "Physique and Character," 1925, 1951. p. 3.
3. Reprinted from Table 2 of Kretschmer [1921] by Sheldon (1963) Sheldon writes that the English edition of Kretschmer's work "contains a number of errors." The nature of these errors is not further specified.
4. Ellenberger, H. F., "The Discovery of the Unconscious: The History and Evolution of Dynamic Psychiatry," 1970.
5-11. Sheldon, W. S. Sheldon, W. H. "(with the collaboration of S. S. Stevens and W. B.Tucker). The Varieties of Human Physique: An Introduction to Constitutional Psychology. New York: Harper, 1940. p. 31, 72, 73, 12, 21. 9 The discussion of the Elgin study is to be found in Hartl, 1982, pp. 26-27 and 35. Reference is made to a 'detailed write up by Phyllis Wittman (1948)'. This would appear to be Wittman, P., Sheldon, W. H., and Katz, C. J. "A study of the relationship between constitutional variations and fundamental psychotic behavior reacxtions," Journal of Nervous and Mental Diseases, 1948, 108, 470-476.
12. Hartl, E. M., Monnelly, E. P. and Elderkin, R. D. "Physique and Delinquent Behavior," 1982, p.44.
13. Hartl, E. M., Monnelly, E. P. and Elderkin, R. D. "Physique and Delinquent Behavior," 1982, p.44.

14. Hartl, E. M., Monnelly, E. P. and Elderkin, R. D. "Physique and Delinquent Behavior," 1982, p.12.
15. Hartl, E. M., Monnelly, E. P. and Elderkin, R. D. "Physique and Delinquent Behavior," 1982, p.12.
16. Hartl, E. M., Monnelly, E. P. and Elderkin, R. D. "Physique and Delinquent Behavior," 1982, p.12.
14. Hartl, E. M., Monnelly, E. P. and Elderkin, R. D. "Physique and Delinquent Behavior," 1982, p.12.
15. Hartl, E. M., Monnelly, E. P. and Elderkin, R. D. "Physique and Delinquent Behavior," 1982, p. 2.
16. Hartl, E. M., Monnelly, E. P. and Elderkin, R. D. "Physique and Delinquent Behavior," 1982, p. 49.
17. Hartl, E. M., Monnelly, E. P. and Elderkin, R. D. "Physique and Delinquent Behavior,' 1982, p. 50.
18. Hartl, E. M., Monnelly, E. P. and Elderkin, R. D. "Physique and Delinquent Behavior," 1982, p. 51.
19. Hartl, E. M., Monnelly, E. P. and Elderkin, R. D. "Physique and Delinquent Behavior," 1982, p. 503.
20-23. Hartl, E. M., Monnelly, E. P. and Elderkin, R. D. "Physique and Delinquent Behavior," 1982, p. 512.
21. Zusne, L. "Biographical Dictionary of Psychology", 1984.

Chapter 7: The Mental Fossil of Predicting Criminality

1. Vold, G. "Theoretical Criminology," 1958. I find this book to be a classic work because of its clarity and thoughtfulness. The first set of topics presented in Chapter 7 follows the pattern Vold establishes in Part II of his book, partly because of the inherent logic in this order of presentation and partly because the choice or order is informative about the shifting ideas I have called mental fossils.
2. Vold, G. "Theoretical Criminology," 1958, p. 109-110.
3. Torrey, E. F., "Out of the Shadows,"1996, 1997.

4. Sokal, M. M. "James McKeen Cattell and Mental Anthropometry: Nineteenth-Century Science and Reform and the Origins of Psychological Testing," Chapter 2 in Sokal, M. M. "Psychological Testing and American Society, 1890-1930," 1987.
5. Sokal, M. M. "James McKeen Cattell and Mental Anthropometry: Nineteenth- Century Science and Reform and the Origins of Psychological Testing," Chapter 2 in Sokal, M. M. "Psychological Testing and American Society, 1890-1930," 1987.
6. The small quotation is from Lombroso, C., "Criminal Man," 1911, p. iv, this prepared by his daughter after his death. It is a summary of the life-work of Lombroso and is remarkable for doing so without hagiography. The book is of value to the English-speaking reader, as much of the original works remain untranslated from the Italian. The quotations are from all from Lombroso, C., "Criminal Mind," 1911, xiv-xvi. And Lombroso, C. "Criminal Mind," 1911, pp 7-16.
7. Vold, G. "Theoretical Criminology," 1958.
8. Goring, C. "The English Convict," 1913. (n.s.)The quotation is from the one edited by Vold, 1958, p. 58.
9. Hooton, E. A. "The American Criminal, volume 1, The Native White Criminal of Native Parentage,"1939, p xi.
10. Hooton, E. A., "Up from the Apes, 1931," "Apes, Men and Morons," 1937. "Man's Poor Relations," 1942; "Up from the Apes," 1931.î
11. Hooton, E. A. "The American Criminal," 1939.
12. Hooton, E. A., "The American Criminal," 1939, p. 306.
13. Hooton, E. A., "The American Criminal," 1939, p. 306.
14. Hooton, E. A., "The American Criminal," 1939, p. 306.
15. Vold, G., "Theoretical Criminology,"1958, pp. 63-65.
16. Hooton, E. A., "The American Criminal," 1939, p. 306.
17. Candland, D. K., "Feral Children and Clever Animals," 1993.

18. Goring, C. "The English Convict," 1913. (n.s.)The quotation is from the one edited by Vold, 1958, p. 58. The history of the mental testing movement is complex, instructive, and far beside the point of this book. An elegant, scholarly, questioning approach is to be found in Sokal, M. M. "Psychological Testing and American Society, 1890-1930," For information on intelligence and the criminal, I have depended on Vold, G. "Theoretical Criminology," 1958.

18-22. Glueck, S. and Glueck, E. "Physique and Delinquency," 1956, pp 265. pp 219, 226, 270. , pp. 268-273. pp. 268-273. Not all would agree that the X2 is the appropriate statistic for use on percentages, since percentages by definition reduce the variance. "Physique andCharacter" includes a chapter "Explanation of Statistical Method" by Jane Worcesrer that explains the statistical problem inherent in the Gluecks' design. This chapter deserves wider attention, for although written for a 1956 publication, the chapter well describes the statistical problem to be addressed in research of the sort conducted by Hooton and the Gluecks. The cautions expressed appear to be ignored in present-day work.

23. Vold, G., "Theoretical Criminology," 1958, p. 129.

24. Yochelson, S. and Samenow, S, E., "The Criminal Personality: Volume I: A Profile for Change." 1976, p. 80-81. Later fossil ideas regarding criminality have identified separation-trauma, especially separation from mother and aggression, as likely contributors. Presently, 'maladaptive parenting' and 'dysfunctional families' receive credit.

25-40. Yochelson, S. and Samenow, S, E., "The Criminal Personality: Volume I : A Profile for Change." 1976. p. 4. 360.

41-53. Lindner, R. "Stone Walls and Men," 1946. Note 30,

p. 255; 31-256; 32-258; 33-259; 34-260; 35-265; 36-269; 37-174; 38-275; 39-278; 40-187; 41- 3242- 362; 42- 363; 43-367; 44-122. Lindner, R. "Rebel without a Cause," 1944, note 6, p. 487; 46, p. 127;50, p. 138; 47, p. 143; 48, p. 144; 49 p.146; 50p. 194; 51, p.106; 52, p. 284; 53, p. 245; 54, p. 268; 55, p. 288, 56 p. 297.

Chapter 8: 'Human Progress' through Eugenics

As was true for our study of the unconscious in Chapter 3, where it was remarked that Ellenberger's book was of such monumental scholarship that anything new could be but a footnote, so a like acknowledgment must be given to Daniel J. Kevles's "In the Name of Eugenics," to G. R. Searle's work on eugenics in Britain, and M. H. Haller's [1963] work on the American hereditarian attitudes. These, along with readings in the -pre 1935 German literature, are the majors sources on which my analysis of the mental fossil of eugenics is based. Kevles, D. J. "In the Name of Eugenics," 1985, p. 59. Kevles describes how, from 1923 to 1952, it was believed that human beings possessed 48 chromosomes. [Kevles, "In the Name of Eugenics,"
1985, p 238-240.] The reasons for this continuing miscount are themselves an example of a folk-psychology, this among scientists, one expressed either by the technical difficulties in counting or the example of the ëemperorís new clothesí depending on oneís degree of charity. If we can miscount the number of chromosomes [and miss the difference in sex] for thirty years, what are we to believe about other scientific 'facts'?
1. Haller, M. H. "Eugenics, Hereditarian Attidudes in American Thought," 1963, p. 139-140.
2. Leonard Darwin also had views on the use of family

income as a measure of genetic suitability.

3. Washington, P. "Madame Blavatsky's Baboon", 1993, p. 95.

4-6. The history is provided by Kevles, D. J. "In the Name of Eugenics," 1985, Chapter 12, p. 176.

7. Haller, 1963, p. 232]. Regarding psychotics, many antipsychotic medicines now in use, either taken voluntarily or by court-order, remove sexual desire. "Canada: Sterilized in Alberta," The Economist, Novermber 9, 1996, p. 46. no author cited. "Nasty Nordic Eugenics," The Economist, August 30, 1997, p. 36, no author cited. McLaren, A. "Our Own Master Race, Eugenics in Canada, 1885-1945," 1990. Koch, L., "Racehygiene i Danmark", 1996; "Nasty Nordic Eugenics," The Economist, August 30, 1997, p. 36, no author cited. Haller, M. H. "Eugenics, Hereditarian Attitudes in American Thought," 1963, p. 139-140. For Japan, see The New York Times, Northeastern Edition, Thursday, September 18, 1997, p. A12. . Searle, G. R. "Eugenics and Politics in Britain 1900-1914." 1976. p 94-95. Ludwig, E. "Genius and Character," 1927. Kaupen-Haas, H. [Ed.] "Der Griff nach Der Bevlerung," 1986; (Ein Buch der Liebe und der Heimat).

8. Haller, M.H. "Eugenics, Hereditarian Attitudes in American Thought," 1963, p. 141.

9. Haller, M. H. "Eugenics, Hereditarian Attitudes in American Thought," 1963, p. 138, 139.

10. Yerkes, R. M.and LaRue, D. W., Outline of the Study of the Self", 1914.

11. Haller, M. H. "Eugenics, Hereditarian Attitudes in American Thought," 1963, 12. Searle, G. R. "Eugenics and Politics in Britain 1900-1914." 1976. p 94-95.

13-17. Haller, M. H. "Eugenics, Hereditarian Attitudes in American Thought," 1963. 15 Searle, G. R. "Eugenics and Politics in Britain 1900-1914." 1976. San

Francisco Chronicle, August 31, 1996. "Chemical Castration".

Chapter 9: Revisiting Psycho-archeology

There are no footnotes.

SOURCES OF FIGURES

(Attempts have been made to locate copyright holders of all materials. Not all replied. Those from the 19th century are assumed to have exceeded their copyright.

Number of figure

2.1, 2.2, 2.3,2.4: Professor Fowler's reading of Mr. Phillips' skull. Author's copy.

2.5: Three young men; Books, Mechanic, Farmer
From figs. 188, 189, 190: Sizer and Drayton, 1889, p. 185.

2.6: Murals from the American Museum of Natural History from the pictures of the American Museum of Natural History by Charles R. Knight. Description is in Freeland and Adams. Murals are copies from Rainger, 1991. © requested from Rainger and University of Alabama Press.

2.7: Faces and body-types, personality, and psychoses, From Kretschmer, 1925.

3.1: The Chicago World's Columbian Exposition of 1893: From Reid Badger, 1979 © requested from Badger and Nelson Hall, follows p 80.

3.2: The Chicago Fair of 1893, The Midway. From Chicago Fair "Glimpses of the World's Fair," No place listed. Laird

& Lee 1893. 44. The fairgoers."Glimpses of the World's Fair, No place listed. Laird & Lee, 1893.

3.3: Photo and caption © requested from New York Times. Photograph by Greg Marinovich. See Daley, New York Times, (Int) Thursday, February 18, 1996.

3.4: Ota Benga, housed at the Bronx Zoo, with a companion. From Bridges, 1974, 226. © Harper.

3.5: Minik, dressed in his Laborador wear, poses for a publicity photograph while he was living in New York. © requested fromKenn Harper, 1986, 2000. Steerforth Press.

3.6: Minik, enculturated and living with the Wallace family near New York City. © requested from Kenn Harper, 1986, 2000. Steerforth Press.

3.7: Alfred Louis Krueger, documentor and friend of Ishi, with Ishi, here dressed in the 'civilized' style. Picture is cropped. © requested from Kroeber, 1961; University of California Press, follows p 100.

3.8: Ishi at work in civilized society, here demonstrating the use of the bow and arrow. © requested from T. Kroeber and University of California Press, 1962, follows p.100.

3.9: Sample exhibit at the Chicago Fair of 1893 on facial types and criminality. From the exhibition of, perhaps, psychology or anthropology. From Brown, 1994. © requested from Brown and University of Arizona Press, 1994.

3.10: In the 1890s, incoming students at some collegeswere photographed (a relatively new technology) in order to measure physical variation. A generation or two later, the same procedure was used to correlate physical variation with temperament, criminality, and personality traits. © requested from Harvard University Press, Ambler and Banta (Eds) 1989.

3.11: The perfect man and perfect women, as calculated as measurements of students at a university, displayed at the Chicago Fair, 1893. From Brown, 1994, p. 45. © requested from Brown and University of Arizona Press.

4.1: Six thinkers: a. Mesmer, from Ellenberger, 1970, facing 330b, Puseygur, Ellenberger, inset beyond 330, c. Gustav Fechner, d. Rudolph Steiner, e. Emanuel Swedenborg, f. Annie Besant © Ellenberger, 1972.

5.1: Motor Centre in the Monkey's Brain. From Wells, 1889.

5.2: Four men, four temperaments; as based on their facial phenotype. From Sizer and Drayton, 1889, p 19 and 20. p

5.3 Types of noses From Wells, New Physiognomy 1867, p 219.

5.4 : Derivatives from classic phrenology found various body parts to be predictors of talent, temperament, and behavior. From Simms, 1887, p. 153 and p. 163.

6.1: also from Kretschmer, 1922, p. 10.

6.2, 6.3, 6.4: all from Sheldon, 1949, pp. 244 and 245; 489-491; 626-628.
Material on boys thirty years later is from Hartl, et al., 1982. Abstracted From pages 140 (number 44); P 317, [no 125], and p 41. Textual material granted only. © from Hartl and Academic Press.

Figure 7.1 From Lombroso, 1911, p 56.

Figure 7.2 from Glueck and Glueck, 1959, p 166-167

Figure 7.3. From Glueck and Glueck, 1959, p. 102, 103.

The interpolated and abstracted quotations on pp 183-188 are from Lindner, 1944.

Figure 8.1: From Kaupen-Haas, 1989 ©requested from Kaupen-Haas and Greno.

Figure 8.2: From Yerkes and LaRue, 1914, p. 17-18.

Figure 8.3: From Goddard, 1911, p. 3.

Figure 8.4: Carrie Buck and husband: From Smith and Nelson 1989.
© Smith and New Horizon Press, 3

INDEX

In e-book presentation, indexed words are linked to the text and there colored or underlined thereby distracting the reader. An additional problem is that cross-referencing or indexing of concepts becomes unworkable.

An index with cross-referencing, concepts, and pagination has been prepared for the reader wishing to utilize a complete index. Please write to the author (dcandlan@bucknell.edu) for a copy.

AUTHOR'S ACKNOWLEDGMENTS

The figures, being taken from 19th century sources, required the definition now available. I am grateful to Kevin Candland, Kevin Candland Studio of San Francisco, for reworking the photographs pixel by pixel.

Although attempts were made to gain copyright permission for all photographs and quotations, not all publishers still exist and of those who do, not all replied to my requests. I have made an earnest attempt to secure copyright permission and thank those who replied.

The contribution to this work by Geoffrey Mayes requires immediate notation and praise. His editing was of the highest order of scholarship and his ability to see what I wanted to say rather than what I sometimes wrote gave clarity and shape to this work. From him I learned much about my own mental fossils and how they interfered from time to time with my expression. It is a special gift to an author to have with me on this journey a talented, intuitive, and thoughtful editor.

I am grateful for the good fortune of having Barry Nazar comment in depth on the ideas here presented. He challenged and extended my thinking at every turn, but especially in regard to the ways in which ideas can be understood to evolve. Once my student, in time I became his.

Others made comments that aided the exposition, perhaps in ways they do not now remember. Thanks to

Marc Hauser, Michael Pereira, William Clements, Donald, Dewsbury, Sarah Bush, Katherine Nowak, Alex Piel, Jerry Eberhart, Kurt Hoffman, Carl Johnson, Steven Horowitz, Pat Ramsay, Will George, Eric Charles, John Stinchcombe, Jean-Francois Cazorla, Christopher Williams, Aaron Wallisch, Michael Gumert, Brian Smith, Calvin Cyphert, Brian Kramer, Gustavo de Melo Lacerda, Sang Lee, Dhaval Vyas, John Papadakis, Stephan Ralescue, Adam Saunders, Karen Seaver, Andrew Smith, Sam Gosling, Sheldon Zedeck, Irving Zucker, Steven Glickman, Donald Riley, Mark Rosenzweig, Carolyn Scott, Dachner Keltner, Richard Lazarus, Mary Lee Jensen, Jon Jensen, David Fletcher, and Pauline Fletcher. Special thanks also to Nancy and Jim Ream, Caryl, Peter, and Max Mezey, and Paul and Ann Homrighausen for the loan of their homes and hospitality.

Certain institutions, whose life is built by attentive folk, allowed me to use their resources and treasures. Chief among these is the Ellen Clarke Bertrand Library of Bucknell University [and its splendid staff], the Harold Washington Center, Chicago, the University of California at Berkeley libraries; especially, the Bancroft Library of that institution.

www.ingramcontent.com/pod-product-compliance
Lightning Source LLC
Chambersburg PA
CBHW070950200526
45161CB00001BA/56